Elementary Geometry of Algebraic Curves: an Undergraduate Introduction

This is a genuine introduction to plane algebraic curves from a geometric viewpoint, designed as a first text for undergraduates in mathematics, or for postgraduate and research workers in the engineering and physical sciences. It is well illustrated, and contains several hundred worked examples and exercises, making it suitable for adoption as a course text. From the familiar lines and conics of elementary geometry the reader proceeds to general curves in the real affine plane, with excursions to more general fields to illustrate applications, such as number theory. By adding points at infinity the affine plane is extended to the projective plane, yielding a natural setting for curves and providing a flood of illumination into the underlying geometry. A minimal amount of algebra leads to the famous theorem of Bézout, whilst the ideas of linear systems are used to discuss the classical group structure on the cubic.

C. G. Gibson is Senior Fellow in Mathematical Sciences, University of Liverpool.

Elementary Geometry of Algebraic Curves: an Undergraduate Introduction

C. G. Gibson

CAMBRIDGE UNIVERSITY PRESS
Cambridge, New York, Melbourne, Madrid, Cape Town, Singapore, São Paulo

Cambridge University Press
The Edinburgh Building, Cambridge CB2 2RU, UK

Published in the United States of America by Cambridge University Press, New York

www.cambridge.org
Information on this title: www.cambridge.org/9780521641401

First published 1998
Reprinted 2001

A catalogue record for this publication is available from the British Library

Library of Congress Cataloguing in Publication data

Gibson, Christopher G., 1940– .
Elementary geometry of algebraic curves / C. G. Gibson.
p. cm.
ISBN 0 521 64140 3 (hardbound)
1. Curves, Algebraic. I. Title.
QA565.G5 1998
516.3′52–dc21 98-36910 CIP

ISBN-13 978-0-521-64140-1 hardback
ISBN-10 0-521-64140-3 hardback

ISBN-13 978-0-521-64641-3 paperback
ISBN-10 0-521-64641-3 paperback

Transferred to digital printing 2006

To Dorle

Contents

List of Illustrations

List of Tables

Preface

For some time I have felt there is a good case for raising the profile of undergraduate geometry. The case can be argued on *academic* grounds alone. Geometry represents a way of thinking within mathematics, quite distinct from algebra and analysis, and so offers a fresh perspective on the subject. It can also be argued on purely *practical* grounds. My experience is that there is a measure of concern in various practical disciplines where geometry plays a substantial role (engineering science for instance) that their students no longer receive a basic geometric training. And thirdly, it can be argued on *psychological* grounds. Few would deny that substantial areas of mathematics fail to excite student interest: yet there are many students attracted to geometry by its sheer visual content. The decline in undergraduate geometry is a bit of a mystery. It probably has something to do with the fashion for formalism which seemed to permeate mathematics some decades ago. But things are changing. The enormous progress made in studying non-linear phenomena by geometrical methods has certainly revived interest in geometry. And for material reasons, tertiary institutions are ever more conscious of the need to offer their students more attractive courses.

0.1 General Background

I first became involved in the teaching of geometry about twenty years ago, when my department introduced an optional second year course on the geometry of plane curves, partly to redress the imbalance in the teaching of the subject. It was mildly revolutionary, since it went back to an earlier set of precepts where the differential and algebraic geometry of curves were pursued simultaneously, to their mutual advantage.

In the final year of study, students could pursue this kind of geometry by following traditional courses on the differential geometry of curves and surfaces. But in the area of algebraic geometry, matters were more problematic. A course on the geometry of algebraic curves seemed to me to be the obvious kind of development. The problem was a dearth of suitable texts. Some had developed from courses lasting for a whole session, where it was possible to attain some distance. By contrast, I was faced with a single semester course, offered over a period which saw a decline in the technical accomplishments of our students. I simply could not hope to be so ambitious. Also I find myself out of sympathy with colleagues who fret that they fail to reach significant results. I belong firmly to the school of thought which believes that it is far better to obtain a thorough appreciation of the basics than to reach some technical pinnacle. Elementary facts (for instance, the fact that the centre of a circle can be defined projectively, rather than metrically) can have a stunning impact on students. My view is that the few who wish to pursue more advanced aspects of the subject can always proceed to higher degrees where their needs will be met.

This book arose from my lecture notes after several years of experimentation. It has gained enormously from the reaction of my students over the years; they have proved to be my harshest critics, and my most helpful advisers, and I owe them a great deal.

0.2 Required Mathematical Knowledge

The intending reader will probably want to know how much mathematical knowledge is assumed of him. Let me first state quite clearly that one of my objectives was to make this book as accessible as possible. I am well aware of the needs of workers in other fields who do not have a substantial mathematical background: and I feel strongly that it is the very beginnings of the subject which need proper exposition. The more experienced reader will find that this book can be viewed as a stepping stone to many excellent texts which assume a higher level of mathematical preparation. In the area of algebra, the most basic requirement is a good understanding of the elements of linear algebra. The abstract concepts of group, domain and field do occur, and are recalled in Section 2.2: but they only occur in a fairly marginal way – you certainly should not be put off just because you are not familiar with these ideas. More substantially, much of the material rests on the unique factorization of polynomials in several variables: however, all the necessary definitions are given, the result itself is carefully stated, and references are given to the proof for those who wish to see it. In the area of analysis, I do assume the elements of calculus in several variables; basically, you need to be able to work out partial derivatives. The reader should be fluent in handling complex numbers, particularly complex roots of unity which appear in many of the calculations. Beyond that I only assume that the reader has come across the Fundamental Theorem of Algebra, i.e. the statement that every polynomial of positive degree in a single variable has at least one complex zero, but you only need the statement of the result. As to geometry, it would certainly help to have a little background (some familiarity with lines and conics for instance) but effectively the book is quite self-contained. I made a conscious decision to make the material independent of virtually any knowledge of topology. In practice that means that a small number of statements are made without proof. More regrettably, that decision precluded the possibility of developing one of the great historical ideas of the subject, that complex curves can be viewed as real surfaces.

0.3 Concerning the Structure

Concerning the structure of the book, I should say that roughly the first half is devoted to curves in the (familiar) affine plane, and the second to curves in the (less familiar) projective plane. I wanted my reader to feel quite comfortable with the mechanics of handling affine curves before making the conceptually difficult transition to the projective plane. One of the main functions of this book is to place algebraic curves in their natural setting (the complex projective plane) where their structure is more transparent. For some readers, particularly those whose background is not in mathematics, this may prove to be a psychological barrier. I can only assure such readers that the reward is much greater than the mental effort involved. History has shown that placing algebraic curves in a natural setting provides a flood of illumination, enabling one much better to comprehend the features one meets in everyday applications. I made a deliberate effort to keep the individual chapters fairly short, adopting the theory that each chapter revolves around one new idea; likewise the sections are brief, and punctuated by a series of 'examples' illustrating the concepts. I have included a collection of exercises, designed to illustrate (and even amplify) the small amount of theory. Each chapter contains sets of exercises, each appearing immediately after the relevant section. I felt it was a service to the mathematics community to gather together a coherent set of exercises for the benefit of teachers; many have been culled from the older literature.

0.4 Concerning the Content

The content of the book is largely classical. There is a tendency in the subject to overemphasize examples of curves drawn from the distant past. I wanted to make the point that the resurgence of geometry is based on the role it plays in the increasing mathematization of the physical sciences. Thus I have indulged my own passion for the curves which arise in engineering kinematics, a sadly neglected subject (the real beginnings of theoretical robotics) which deserves to be better known both for its intrinsic interest and its considerable mathematical potential. I make no apologies for the fact that conics occupy a substantial part of the text; they play a significant role in geometry at this level, and my view is that their intrinsic importance should be reflected in the space devoted to them. On the same basis, cubics receive an extended discussion. In particular, I regard the group structure on the cubic as one of the most attractive topics of elementary geometry within the reach of a mathematics undergraduate; for me, it is the mathematical equivalent of a treasured holiday snapshot. So far as objectives are concerned, I felt it was sensible to get as far as Bézout's Theorem, to justify in some measure the assertion that algebraic curves live naturally in the complex projective plane.

Some topics are conspicuous by their absence. For instance, I have great affection for the lost art of tracing algebraic curves, to which Frost's classic text on 'Curve Tracing' is a fitting memorial. Like archery, it is a satisfying pursuit, of little relevance to the world we live in. But just as the machine gun has rendered the bow obsolete, so the computer has proved itself a superbly efficient tool for tracing curves at phenomenal speeds. In this connexion, I am particularly grateful to Wendy Hawes, who constructed a picturebook of algebraic curves using the graphics facilities of the Pure Mathematics Department in The University of Liverpool, and kindly allowed me to include her pictures. Finally, I offer my warmest thanks to my friend and colleague Bill Bruce who read a working draft, and produced a wealth of valuable comment.

1
Real Algebraic Curves

Plane curves arise naturally in numerous areas of the physical sciences (such as particle physics, engineering robotics and geometric optics) and within areas of pure mathematics itself (such as number theory, complex analysis and differential equations). In this introductory chapter, we will motivate some of the basic ideas and set up the underlying language of affine algebraic curves. That will also give us the opportunity to preview some of the material you will meet in the later chapters.

1.1 Parametrized and Implicit Curves

At root there are two ways in which a curve in the real plane $\mathbb{R}^2$ may be described. The distinction is quite fundamental.

- A curve may be defined *parametrically*, in the form $x = x(t)$, $y = y(t)$. The parametrization gives this image a dynamic structure: indeed at any parameter value t we have a *tangent vector* $(x'(t), y'(t))$ whose length is the *speed* of the curve at the parameter t. An example is the line parametrized by $x = t$, $y = t$, with constant speed $\sqrt{2}$, another parametrization such as $x = 2t$, $y = 2t$ yields the same image, but at twice the speed $2\sqrt{2}$.
- A curve may be defined *implicitly*, as the set of points (x, y) in the plane satisfying an equation $f(x, y) = 0$, where $f(x, y)$ is some reasonable function of x, y. For instance the line parametrized by $x = t$, $y = t$ arises from the function $f(x, y) = y - x$. Such a curve has no associated dynamic structure – it is simply a set of points in the plane.

Broadly speaking, the study of parametrized curves represents the beginnings of a major area of mathematics called *differential geometry*, whilst the study of curves defined implicitly represents the beginnings of another major area, *algebraic geometry*. It is the latter study which provides the material for this book, though at various junctures we will have something to say about the question of parametrization.

The common feature of many curves which appear in practice is that they are defined implicitly by equations of the form $f(x, y) = 0$ where $f(x, y)$ is a *real polynomial* in the variables x, y, i.e. given by a formula of the shape

$$f(x, y) = \sum_{i,j} a_{ij} x^i y^j$$

where the sum is finite and the coefficients a_{ij} are real numbers. There is much to gain in restricting attention to such curves, since they enjoy a number of important 'finiteness' properties. Moreover, it will be both profitable and illuminating to extend the concepts to situations where the coefficients a_{ij} lie in a more general 'ground field'. In some sense the complexity of a polynomial $f(x, y)$ is measured by its *degree*, i.e. the maximal value of $i + j$ over the indices i, j with $a_{ij} \neq 0$. Given a polynomial $f(x, y)$ we define its *zero set* to be

$$V_f = \{(x, y) \in \mathbb{R}^2 : f(x, y) = 0\}.$$

Instead of saying that a point (x, y) lies in the zero set of a curve f we may, for linguistic variety, say that (x, y) lies on the curve f, or that f passes through (x, y). Note that the zero set (and the degree) are unchanged when we multiply f by a non-zero scalar. It is for that reason that we introduce the following formal definition. A *real algebraic curve* is a non-zero real polynomial f, *up to multiplication by a non-zero scalar*. The more formally inclined reader may prefer to phrase this in terms of 'equivalence relations'. Two polynomials f, g are *equivalent*, written $f \sim g$, when there exists a non-zero scalar λ for which $g = \lambda f$. It is then trivially verified that $\sim$ has the defining properties of an equivalence relation: it is *reflexive* ($f \sim f$), it is *symmetric* (if $f \sim g$ then $g \sim f$), and it is *transitive* (if $f \sim g$ and $g \sim h$ then $f \sim h$). A real algebraic curve is then formally defined to be an equivalence class of polynomials under the relation $\sim$. So strictly speaking, a real algebraic curve is an equivalence set of all polynomials $\lambda f(x, y)$ with $\lambda \neq 0$, and any polynomial in this set is a *representative* for the curve. In this book we will usually abbreviate the term 'algebraic curve' to 'curve'. Curves of degree 1, 2, 3, 4, ... are called *lines*, *conics*, *cubics*, *quartics*, ... It is a long established convention that the curve with representative polynomial $f(x, y)$ is referred to as the 'curve' $f(x, y) = 0$. There is no harm in this provided you remember that it is a convention, and not a shorthand for the zero set. Thus $y = x$ and $2y = 2x$ represent the same curve of degree 1.

It is an unfortunate fact of life that when dealing with the simplest possible curves of elementary geometry (such as the lines and standard conics discussed in the next section) the distinction between curves and their zero sets can be blurred without undue consequences. However, as one proceeds into algebraic geometry the relation between the two concepts becomes crucial, and leads to some of the most fundamental results in the subject. The reader is warned, even at this very early stage, to make a crystal clear mental distinction between the concept of a curve, and that of its zero set.

1.2 Introductory Examples

In this section we present a small selection of curves, illustrating some of the general concepts which will occur later. For reasons of space, it is simply not feasible to give an account of even the more significant situations (in the physical sciences, and within pure mathematics itself) where curves arise, as each such situation would demand at least some of the pertinent underlying mathematics to be developed. However, the impatient reader, wishing to see 'real' curves (in the sense of 'real' ale), might like to jump to Section 1.3 which presents some of the curves arising in planar kinematics. A good guiding philosophy is to begin at the beginning (though we will not end at the end) and work with increasing degree. Curves of degree 1 are lines, and play a fundamental role in understanding the geometry of general curves. We will recall their most important attributes via a series of examples. According to the above definition a line has the form $ax + by + c$ with at least one of a, b non-zero.

Example 1.1 Given any two distinct points $p = (p_1, p_2)$, $q = (q_1, q_2)$ in $\mathbb{R}^2$ there is a unique line $ax+by+c$ passing through p, q. We seek scalars a, b, c (not all zero) for which

$$ap_1 + bp_2 + c = 0, \qquad aq_1 + bq_2 + c = 0.$$

Since p, q are distinct, the 2×3 coefficient matrix of these two linear equations in a, b, c has rank 2. By linear algebra it has kernel rank 1, so there is a non-trivial solution (a, b, c), and any other solution is a non-zero scalar multiple of this one. Explicitly, the line joining p, q is given by

$$(p_1 - q_1)(y - p_2) = (p_2 - q_2)(x - p_1).$$

Note this point well: *the equation of a line is determined up to scalar multiples by its zero set.* For higher degree curves that can fail.

Example 1.2 Consider two lines $a_1x + b_1y + c_1$, $a_2x + b_2y + c_2$ in $\mathbb{R}^2$. The intersection points (x, y) are those points which satisfy the linear equations

$$a_1x + b_1y + c_1 = 0, \qquad a_2x + b_2y + c_2 = 0.$$

By linear algebra, when the determinant $\delta = a_1b_2 - a_2b_1 \neq 0$ these equations have a *unique* solution (x, y). Otherwise, there is no solution (*parallel lines*) or a line of solutions (*coincident lines*).

Example 1.3 Let l be a line, and let p be a point not on l. Then there is a unique line m through p parallel to l. Suppose that l has equation $ax + by + c = 0$. It follows from the previous example that the lines m parallel to l are those of the form $ax + by + d = 0$, with d arbitrary. The condition for m to pass through p then determines d uniquely.

Example 1.4 Lines can be parametrized in a natural way. Consider a line $ax + by + c$, and *distinct* points $p = (p_1, p_2)$, $q = (q_1, q_2)$ on the line. Then a brief calculation verifies that any point $p + t(q - p) = (1 - t)p + tq$ also lies on the line. Conversely, we claim that any point $r = (r_1, r_2)$ on the line has the form $r = (1 - t)p + tq$ for some scalar t. Since p, q, r all lie on the line we have

$$\left\{ \begin{array}{lcl} ap_1 + bp_2 + c & = & 0 \\ aq_1 + bq_2 + c & = & 0 \\ ar_1 + br_2 + c & = & 0. \end{array} \right.$$

That is a linear system of three equations in a, b, c. Since at least one of a, b is non-zero, the system has a non-trivial solution. By linear algebra, the 3×3 matrix of coefficients is singular, so the rows $(p_1, p_2, 1)$, $(q_1, q_2, 1)$, $(r_1, r_2, 1)$ are linearly dependent. However, the first two rows are linearly independent (as p, q are distinct) so the third row is a linear combination of the first two, i.e. $(r_1, r_2, 1) = s(p_1, p_2, 1) + t(q_1, q_2, 1)$ for some scalars s, t. That means $r = sp + tq$ and $1 = s + t$, so $r = (1 - t)p + tq$, as required.

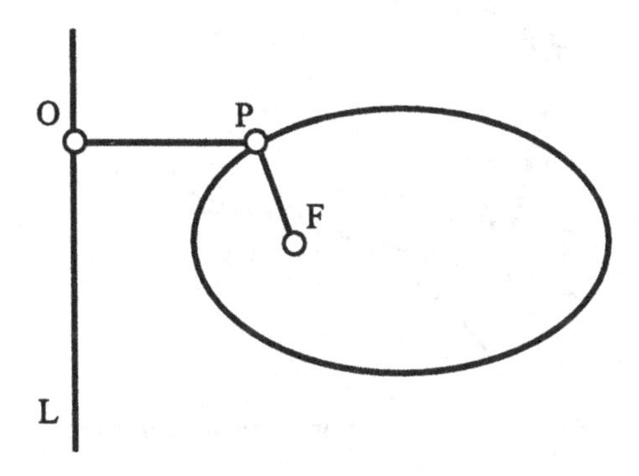

Fig. 1.1. Constructions of standard conics

Note that the parametrization of a line depends on the choice of points p, q. Given two distinct points $p = (p_1, p_2)$, $q = (q_1, q_2)$, the *parametrized* line through p, q is the specific parametric curve given by the above example, namely $x = (1-t)p_1 + tq_1$, $y = (1-t)p_2 + tq_2$.

Example 1.5 Conics will be a recurrent theme in this text. The most familiar conic by far is the circle, defined metrically as the locus of points (x, y) whose distance from a fixed point (a, b) in the plane (the *centre*) takes a constant value $r > 0$ (the *radius*). A circle is thus the zero set of a polynomial $(x-a)^2 + (y-b)^2 = r^2$. The 'standard' parametrization of the circle is $x = a + r\cos t$, $y = b + r\sin t$, but we will meet other parametrizations later.

Example 1.6 The reader has probably met the 'standard conics' of elementary geometry via a metrical construction going back to the classical Greeks. One is given a line L (the *directrix*), a point F (the *focus*) not on L, and a variable point P whose distance from F is proportional to its distance from L. O denotes the unique point on L for which L is perpendicular to the line through O, F. (See Figure 1.1.)

The locus of P is known as a 'parabola', an 'ellipse', or a 'hyperbola' according as the constant of proportionality e (the *eccentricity*) is $= 1$, < 1 or > 1. The fact that P lies on a conic is demonstrated by taking O to be the origin, L to be the y-axis, and the line OF to be the x-axis (with F on the positive axis). Then, setting $F = (2a, 0)$ with $a > 0$, the condition on $P = (x, y)$ is $(x-2a)^2+y^2 = e^2x^2$, which is indeed a conic. For instance in the case $e = 1$ of a parabola this becomes $4a(a-x) + y^2 = 0$: the translation $x = X+a$, $y = Y$ then yields the *standard parabola* $Y^2 = 4aX$

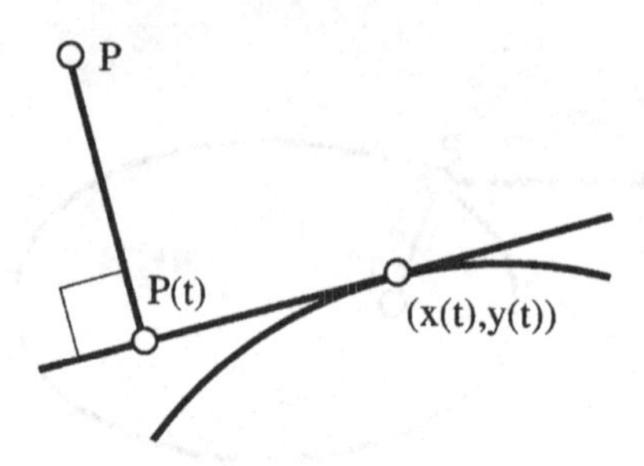

Fig. 1.2. Construction of pedal curves

with focus $F = (a, 0)$ and directrix $X = -a$. In Chapter 4 we will have more to say about the process of reducing polynomials to such 'normal' forms by applying translations, and more generally 'affine mappings'.

Example 1.7 A *regular* parameter of a parametrized curve $x = x(t)$, $y = y(t)$ is a parameter t for which the tangent vector $(x'(t), y'(t))$ is non-zero. The unique line through $P = (x(t), y(t))$ in the direction of the tangent vector is the *tangent line* to the parametrized curve at t, given parametrically as $x = x(t) + \lambda x'(t)$, $y = y(t) + \lambda y'(t)$. It is the line

$$y'(t)x - x'(t)y + \{x'(t)y(t) - x(t)y'(t)\} = 0.$$

Tangent lines play a fundamental role in studying parametrized curves. We will discuss tangent lines to algebraic curves in Chapter 7 and relate them to the concept just introduced for parametrized curves. Numerous interesting constructions are based on the tangent lines to parametrized curves, and give rise to a zoo of interesting curves. One such construction is that of the 'pedal', of considerable importance in geometric optics and kinematics. Suppose we are given a *regular* parametrized curve $x = x(t)$, $y = y(t)$, i.e. one for which every parameter t is regular, and a fixed point $P = (\alpha, \beta)$, the *pedal point*. Then the *pedal* curve of the curve with respect to P is the parametrized curve obtained by associating to the parameter t the projection $P(t)$ of P onto the tangent line at t. (Figure 1.2.)

In practice, given the tangent line, you can write down the line perpendicular to it through P and find the intersection $P(t)$ of the two lines. (Recall from elementary geometry that the lines perpendicular to a given line $ax + by + c = 0$ are the lines of the form $-bx + ay + d = 0$.) Here is a deceptively simple example giving rise to a number of interesting cubics.

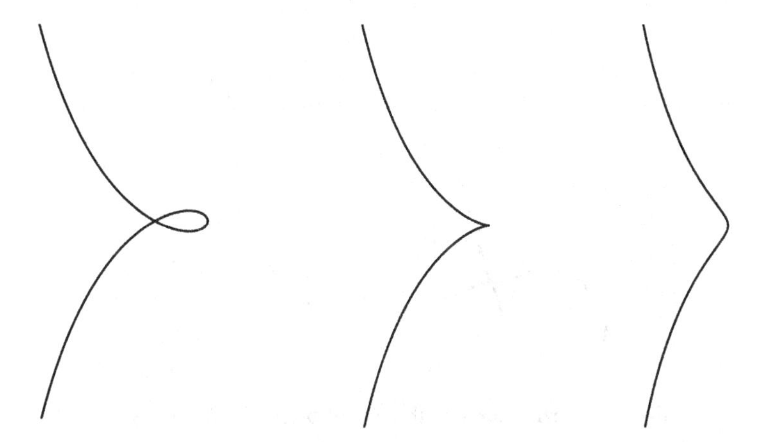

Fig. 1.3. Pedals of a parabola

Example 1.8 Consider the standard parabola $y^2 = 4ax$ with $a > 0$, parametrized as $x = at^2$, $y = 2at$. We will show that the pedal curve with respect to the point $P = (\alpha, 0)$ satisfies the equation of a cubic. The tangent line at t is $x - ty + at^2 = 0$, and the perpendicular line through P is $tx + y - \alpha t = 0$. The parametrized pedal is obtained by setting these expressions equal to zero, and then solving for x, y in terms of t, to obtain

$$x = \frac{(\alpha - a)t^2}{1 + t^2}, \qquad y = \frac{t(\alpha + at^2)}{1 + t^2}.$$

To obtain a polynomial satisfied by the points on the pedal we eliminate t instead, to obtain the cubic $x(x-\alpha)^2 + y^2(a-\alpha+x) = 0$. More precisely we have obtained a *family* of cubics, depending on α. The zero set of some of the pedal curves are illustrated in Figure 1.3.

The first thing to notice is that P always lies on the pedal, and is in some visual sense 'singular'. Thus for $\alpha < 0$ the curve has a loop, which crosses itself at P, for $\alpha = 0$ the loop contracts down to a point, giving a 'cusp' at P, and for $\alpha > 0$ the curve has an isolated point at P. Such 'singular' points play a very basic role in understanding the geometry of a curve, and will be studied in some detail in Chapter 6. The cubic $x^3 + y^2(x + a) = 0$ obtained when $\alpha = 0$ is called the *cissoid of Diocles* after the classical Greek mathematician Diocles, who derived its equation when solving the problem of 'doubling the cube'; Newton discovered a mechanical construction for the cissoid, which we will meet in Section 1.3. $\alpha = 0$ is not the only exceptional value of α. When $\alpha = a$,

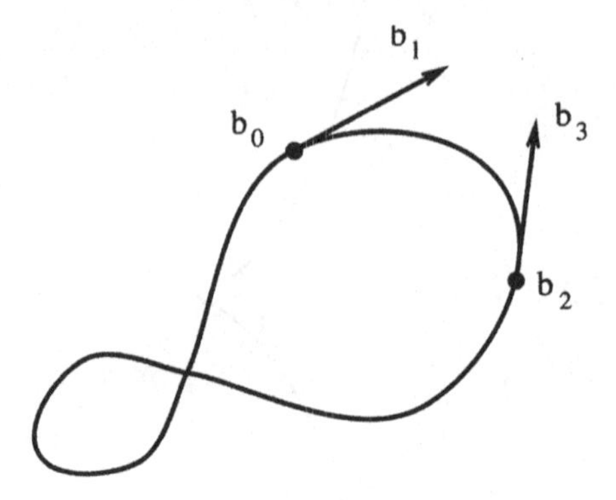

Fig. 1.4. Bézier curve with control points b_0, b_1, b_2, b_3

i.e. the pedal point is the focus F of the parabola, the equation factorizes as $x\{(x-a)^2+y^2\}=0$, whose zero set is simply the y-axis together with the point F. Another exceptional value is $\alpha=-a$, i.e. the pedal point is the point of intersection of the axis and directrix of the parabola: the pedal curve is then known as the *right strophoid*, and is characterized geometrically within the family by the fact that the tangents to the pedal curve at P are perpendicular.

Cubic curves play important roles in numerous areas of mathematics and the physical sciences. An interesting class of naturally parametrized cubics arises in Computer Aided Design (CAD). The idea is as follows. One is given a plane 'curve', for instance part of an artist's visualization of an industrial product, and one seeks a useful mathematical model for this curve which can be handled on a computer. The underlying idea was developed in the late 1950s by two design engineers working for rival French car companies, namely Bézier (working for Renault) and de Casteljau (working for Citröen). A first crude step is to take a sequence of points $b_0, b_2, \ldots, b_{2n}$ on the curve and interpolate a polynomially parametrized curve. However, this process is intrinsically unsatisfactory: as n increases, so the degrees of the polynomials increase, and the interpolating curve may oscillate wildly. The idea is to control this oscillation by specifying the tangent direction at each point. One way of doing this is to associate to each point b_{2k} another point b_{2k+1} and stipulate that the tangent direction of the interpolating curve at b_{2k} should be the direction of the line segment joining b_{2k}, b_{2k+1}. Let us illustrate this for the case of two points b_0, b_2. In that case there are four *control points* b_0, b_1, b_2, b_3 to which is associated the *Bézier curve* defined by

$$B(t) = (1-t)^3 b_0 + 3t(1-t)^2 b_1 + 3t^2(1-t)b_3 + t^3 b_2.$$

Note that $B(0) = b_0$, $B(1) = b_2$, $B'(0) = 3(b_1 - b_0)$, $B'(1) = 3(b_2 - b_3)$ so the curve passes through the points b_0, b_2 and has tangent directions $b_1 - b_0$, $b_2 - b_3$ at those points. (Figure 1.4.) What is not obvious is that *Bézier curves are algebraic*. For the moment we will content ourselves with a numerical example. Later (Example 14.4) when we have a little more algebra available, we will be able to prove this in full generality.

Example 1.9 In the above discussion take $b_0 = (0,0)$, $b_1 = (1/3,0)$, $b_2 = (2,2)$, $b_3 = (1,1/3)$ so $B(t) = (t+t^2, t^2+t^3)$. Write $x = t+t^2$, $y = t^2+t^3$. Note that $y = tx$. Eliminating t we see that each point (x,y) on the Bézier cubic lies on the cubic curve $x^3 = y(x+y)$. Conversely, we will show that any point (x,y) satisfying the equation $x^3 = y(x+y)$ necessarily has the form $x = t+t^2$, $y = t^2+t^3 = tx$ for some real number t. Indeed if $x \neq 0$ *define* t by the relation $y = tx$; then, substituting for y we obtain $x = t+t^2$, and hence $y = tx = t^2+t^3$. Finally, if $x = 0$ then $y = 0$ and we can choose either $t = 0$ or $t = -1$.

Numerous examples of quartic curves arise in the physical sciences. For the moment we will content ourselves with a particularly interesting family of quartics.

Example 1.10 The unit circle $x^2 + y^2 = 1$ is parametrized as $x = \cos t$, $y = \sin t$. We will find the pedal curve with respect to a point $p = (\alpha, 0)$ on the x-axis with $\alpha > 0$. The tangent line at t is $(\cos t)x + (\sin t)y - 1 = 0$, and the perpendicular line through p is $\sin t(x-\alpha) - (\cos t)y = 0$. An equation for the pedal can be found by solving these relations for $\sin t$, $\cos t$, and then substituting in the identity $\cos^2 t + \sin^2 t = 0$. The result is the quartic curve

$$\{x(x-\alpha) + y^2\}^2 = (x-\alpha)^2 + y^2$$

known as a *limacon*. The zero set of the limacon (Figure 1.5) depends on the value of α. The point p always lies on the pedal, and is in some sense 'singular'. (Compare with the pedals of the parabola.) For $\alpha > 1$ the curve has two loops, whilst for $\alpha < 1$ it has just one. The intermediate case $\alpha = 1$ gives rise to a curve with a 'cusp', known as a *cardioid*.

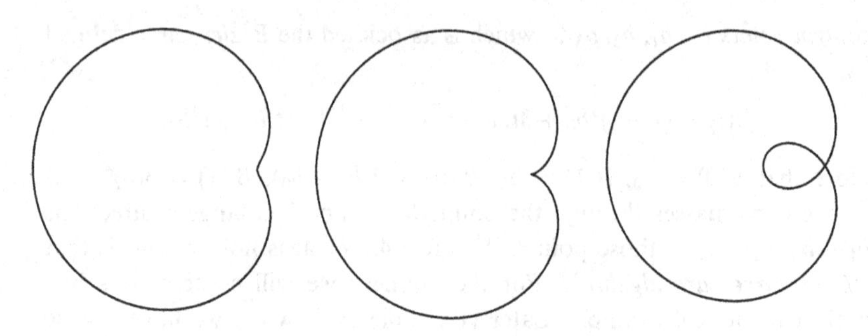

Fig. 1.5. The three forms of a limacon

Exercises

1.2.1 Show that the Bézier curve defined by the points $b_0 = (-9, 0)$, $b_1 = (-9, -1)$, $b_2 = (-6, -2)$, $b_3 = (-8, -2)$ is given parametrically by $x = 3(t^2 - 3)$, $y = t(t^2 - 3)$. By eliminating t from these relations, show that every point lies on the zero set of *Tschirnhausen's cubic* $27y^2 = x^2(x + 9)$.

1.2.2 A parametrized curve is defined by $x(t) = t^2 + t^3$, $y(t) = t^3 + t^4$. Find a polynomial $f(x, y)$ of degree 4 such that $f(x(t), y(t)) = 0$ for all t. (It helps to observe that $y = tx$.) Conversely, show that for any point (x, y) with $f(x, y) = 0$ there exists a real number t with $x = x(t)$, $y = y(t)$. (Again, it helps to observe that you seek a t for which $y = tx$.)

1.2.3 Show that there exists a cubic curve $f(x, y)$ such that every point on the parametrized curve $x(t) = 1 + t^2$, $y(t) = t + t^3$ satisfies the equation $f(x(t), y(t)) = 0$. Conversely, show that for any point (x, y) with $f(x, y) = 0$, *with one exception*, there exists a real number t with $x = x(t)$, $y = y(t)$.

1.2.4 Let $a > 0$, let C be the circle of radius a with centre $(a, 0)$ and let D be the line $x = 2a$. For each line L through the origin O (except the y-axis) let C_L, D_L denote the points where L meets C, D respectively, and let B_L denote the point on the line segment joining O, D_L for which $OB_L = C_L D_L$. Taking t to be the angle between L and the x-axis, find the coordinates of C_L, D_L in terms of t, and hence show that the locus of B_L has the parametric form $x(t) = 2a \sin^2 t$, $y(t) = 2a \sin^2 t \tan t$ with $-\pi/2 < t < \pi/2$. Verify that every point on this parametrized curve lies on the zero set of the cubic $x^3 = (2a - x)y^2$. Conversely, show that any point (x, y) on this cubic curve satisfies the relation $0 \leq x < 2a$, and deduce that it has the form $x = x(t)$, $y = y(t)$ for some t with $-\pi/2 < t < \pi/2$. (The cubic in this question is the cissoid of Diocles, obtained via its original construction.)

1.2.5 Show that there exists a cubic curve $f(x, y)$ such that every point on the parametrized curve $x(t) = \sin 2t$, $y(t) = \sin 2t \tan t$ satisfies $f(x(t), y(t)) = 0$. Conversely, show that for every point (x, y) with $(x, y) = 0$ there exists a t for which $x = x(t)$, $y = y(t)$.

1.2.6 Let $a > 0$. A parametrized curve is defined by $x(t) = 2a \tan t$, $y(t) = 2a \cos^2 t$ where $-\pi/2 < t < \pi/2$. Show that there exists a cubic curve $f(x, y)$ such that $f(x(t), y(t)) = 0$ for all parameters t. Let (x, y) be a point with $f(x, y) = 0$. Show that $0 < y \leq 2a$. Use this fact to show that there exists a parameter t for which $x = x(t)$ and $y = y(t)$. Sketch the zero set of f. (f is the curve known as *Agnesi's versiera*.)

1.2.7 Let $a > 0$. A parametrized curve is defined by the formulas

$$x(t) = \frac{2a \cos t}{1 + \sin t}, \quad y(t) = a(1 + \sin t)$$

where $-\pi/2 < t < \pi/2$. Show that there exists a cubic curve $f(x, y)$ such that $f(x(t), y(t)) = 0$ for all t. Let (x, y) be a point with $f(x, y) = 0$. Show that $0 < y \leq 2a$. Use this fact to show that there exists a t for which $x = x(t)$ and $y = y(t)$. Sketch the zero set of f. (f is identical to the curve in Exercise 1.2.6.)

1.2.8 Let $a > 0$. A parametrized curve is defined by $x(t) = a \cos t$, $y(t) = a \sin t \cos t$. Show that there exists a quartic curve $f(x, y)$ such that $f(x(t), y(t)) = 0$ for all values of t. Let (x, y) be a point in the plane with $f(x, y) = 0$. Show that $-a \leq x \leq a$. Use this fact to show that there exists a real number t for which $x = x(t)$ and $y = y(t)$. Sketch the zero set of f. (f is the *eight-curve*.)

1.2.9 Let $a, b > 0$. Every line L through $(0, 0)$ in $\mathbb{R}^2$, except the line $x = 0$, meets the line $x = b$ in a unique point Q_L: and there are two points on L distance a from Q_L. K is the set of all such points. (At this point it helps to draw a sketch to understand the question.) Show that K is parametrized as $x(t) = b + a \cos t$, $y(t) = b \tan t + a \sin t$ with $-\pi/2 < t < \pi/2$ or $\pi/2 < t < 3\pi/2$, explaining what these intervals correspond to. (Take t to be the angle between the line L and the x-axis, and use a little trigonometry.) Find a quartic curve $f(x, y)$ such that every point in K satisfies the equation $f(x, y) = 0$. Conversely, show that any point $(x, y) \neq (0,0)$ on this quartic has the form $x = x(t)$, $y = y(t)$ for some t in one of the given intervals. Show that $(0,0)$ lies on K if and only if $b \leq a$. (f is the curve known as the *conchoid of Nicomedes*.)

1.2.10 Let $a, b > 0$. A parametrized curve is defined by $x(t) = a(1+\cos t)$, $y(t) = b \sin t(1 + \cos t)$. Find a quartic curve f of the form $f(x, y) = y^2 - \phi(x)$ where $\phi(x)$ is a quartic in x, with the property that $f(x(t), y(t)) = 0$ for all t. Let (x, y) be a point with $f(x, y) = 0$. Show that $0 \leq x \leq 2a$. Use this fact to show that there exists a t with $x = x(t)$ and $y = y(t)$. Give a rough sketch of the zero set of f.

1.2.11 Show that there exists a quartic curve $f(x, y)$ such that every point on the *three-leaved clover* given by $x(t) = \cos 3t \cos t$, $y(t) = \cos 3t \sin t$ satisfies $f(x(t), y(t)) = 0$. Conversely, show that for every point (x, y) with $f(x, y) = 0$ there exists a t for which $x = x(t)$, $y = y(t)$.

1.2.12 A parametrized curve in the real plane $\mathbb{R}^2$ is given by the following formulas, where a, b are positive real numbers, and t is such that $\cos t \neq 0$, $\sin t \neq 0$.

$$x(t) = \frac{a}{\cos t}, \qquad y(t) = \frac{b}{\sin t}.$$

By eliminating t from these relations, show that there exists a quartic curve $f(x, y)$ such that $f(x(t), y(t)) = 0$ for all values of t. Let (x, y) be any point in the plane with $f(x, y) = 0$; show that *either* $(x, y) = (0,0)$ *or* $|a/x| < 1$ and $|b/y| < 1$. Deduce that for any point (x, y) with $f(x, y) = 0$, with one exception, there exists a real number t for which $x = x(t)$ and $y = y(t)$. Sketch the zero set of the curve $f(x, y)$.

1.3 Curves in Planar Kinematics

It is not possible (and probably not desirable) in a book of this length to describe the numerous areas of the physical sciences which contribute to the genesis of curve theory. A compromise is to mention one area of particular historical significance, namely planar kinematics. During the Industrial Revolution, mechanisms for converting rotary into linear motion were widely adopted in industrial and mining machinery, locomotives and metering devices. Such devices had to combine engineering simplicity with a high degree of accuracy, and the ability to operate at

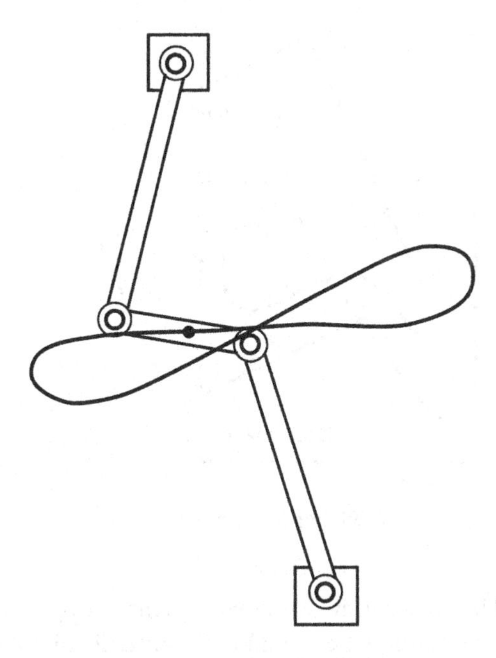

Fig. 1.6. Watt four bar linkage

speed for lengthy periods. For many purposes approximate linear motion is an acceptable substitute for exact linear motion. Perhaps the best known example is the Watt four bar linkage, discovered by the Scottish engineer James Watt in 1784. (Figure 1.6.) The device is made up of three smoothly jointed bars moving with one degree of freedom, the mid-point of the middle (or coupler) bar describing the *Watt curve*. The curve has a self-crossing with two 'branches' through it, one of which gives an excellent approximation to a straight line: it is a particularly good example of the 'flex' concept of Chapter 13.

It was the detailed investigation of the fascinating curves traced by mechanisms such as the Watt four bar which gave rise to the (sadly neglected) body of knowledge known as planar kinematics. The underlying idea can be abstracted as follows. Suppose that a copy of the plane (the *moving plane*, in the language of that subject) moves with one degree of freedom (dof) over the given plane (the *fixed plane*). Any choice of *tracing point* P in the moving plane will then trace a *trajectory* curve on the fixed plane, yielding a 2-parameter family of trajectories, one for each point P. The idea is illustrated in Figure 1.7.

It is worthwhile thinking a little bit about how such motions can

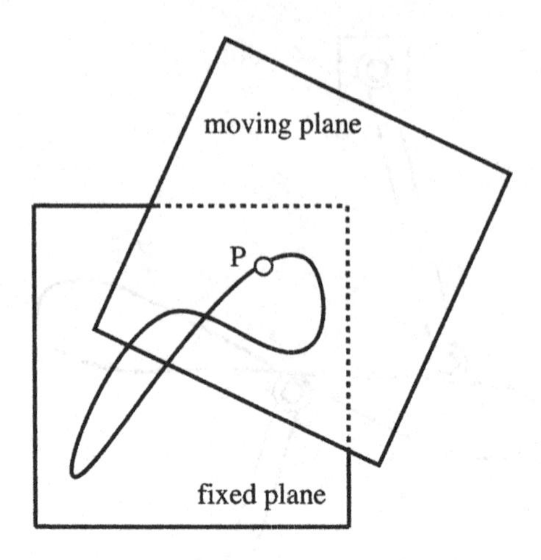

Fig. 1.7. Planar motion with one dof

be generated. The unconstrained moving plane moves with three dof, represented by one rotational dof, and two translational dof. In principle, the movement will be restricted to one dof when the moving plane is subject to two constraints. One way of constraining the motion is to insist that a given point in the moving plane must lie on a given curve in the fixed plane; or dually, one can insist that a given point in the fixed plane must lie on a given curve in the moving plane. Motions with one dof can be constructed by imposing two such constraints, of the same or different types.

- *Construction 1*. Choose points B, C in the moving plane, and curves L, M in the fixed plane, and insist that B lies on L, and that C lies on M.
- *Construction 2*. In the moving plane take a point B and a curve M, in the fixed plane take a point C and a curve L, and insist that B lies on L, and C lies on M.

By making explicit choices of curves L, M one obtains a wide range of planar motions of considerable practical and theoretical interest. Here are some examples.

Example 1.11 The simplest example of a planar motion generated according to Construction 1 is obtained when the two given curves in

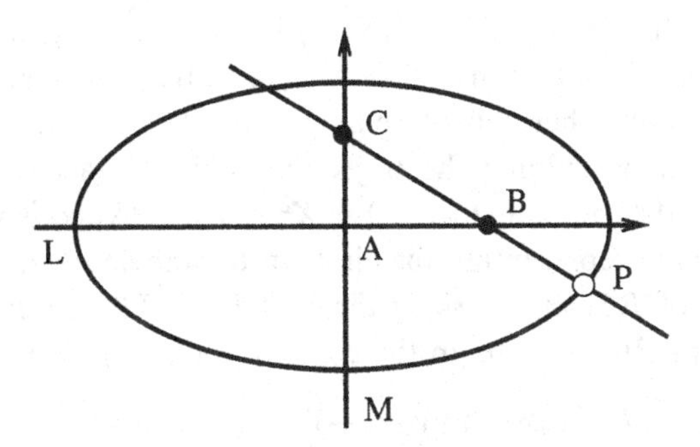

Fig. 1.8. The double slider motion

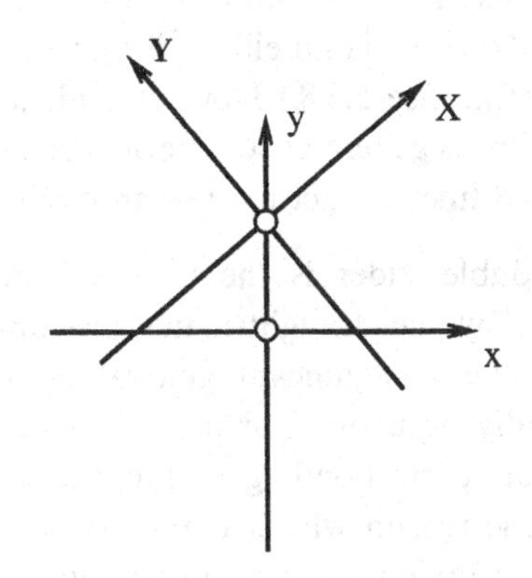

Fig. 1.9. The fixed and moving coordinate systems

the fixed plane are lines L, M, intersecting at a single point A, this is the *double slider* motion of engineering kinematics, illustrated in Figure 1.8.

We will show that *the trajectories of the double slider motion are conics*. For simplicity we assume L, M are perpendicular. In the fixed plane we take the x-, y-axes to be the lines L, M, with A the origin, and in the moving plane we take the X-, Y-axes to be the line L' joining B, C and the line M' obtained by rotating L' through a right angle about C. (Figure 1.9.)

Since B, C are fixed in the moving plane they have to remain a constant distance $d > 0$ apart. It is convenient to parametrize points on L, M as $B = (ds, 0)$, $C = (0, dt)$ so the constant distance condition is $s^2 + t^2 = 1$. The coordinate systems in the fixed and moving planes are then related by

$$x = sX + tY, \quad y = sY + t(d - X). \tag{$\star$}$$

Choose a fixed tracing point $P = (X, Y)$ in the moving plane. Provided the matrix of coefficients is invertible, we can solve these equations for s, t in terms of x, y and then substitute in $s^2+t^2 = 1$ to obtain a polynomial of degree 2 in x, y defining the trajectory of P. In fact the matrix of coefficients has determinant $F(X, Y) = X^2 + Y^2 - dX$, so is invertible if and only if $F = 0$, representing the circle in the moving plane centred at the mid-point of BC, and passing through B, C. Eliminating s, t from (⋆) via Cramer's Rule we obtain the conic in the (x, y)-plane given by

$$((d-X)x - Yy)^2 + (-Yx + Xy)^2 = F^2.$$

More precisely, we have a family of conics, depending on the choice of tracing point $P = (X, Y)$. The nature of the conic depends on this choice. For instance, the conic is an ellipse if and only if P does not lie on the circle $F = 0$. (Exercise 5.2.8.) However, when P does lie on the circle $F = 0$ the equation degenerates to a perfect square $(Yx - Xy)^2 = 0$, representing a 'repeated line', in fact the line through P and the origin.

Incidentally, the double slider is the basis of the mechanical construction of an ellipse by the draughtsman's instrument known as the *ellipsograph*. It illustrates a significant general point in curve theory, namely that in naturally occurring *families* of curves we should expect degenerations to occur, corresponding to factorization of the defining polynomial. That is one reason why it is important to understand how polynomials factorize, a topic we pursue in Chapter 3.

Example 1.12 The simplest instance of a planar motion constructed according to Construction 2 is when both the curves L, M are straight lines. We will call it a *Newton motion*, since it was Newton who realized that the cissoid of Diocles could be constructed as a trajectory of such a motion. Write B' for the point on M with the property that the line BB' is perpendicular to M: and likewise write C' for the point on L with the property that the line CC' is perpendicular to L. For simplicity we suppose the tracing point $P = (x, y)$ is the mid-point of the line segment joining B, B', and we suppose that the distances from B to M, and from C to L, take the fixed value $2a > 0$. (Figure 1.10.) It will be no restriction to assume that $C = (-a, 0)$ and that L is the line $x = a$. Since B lies on L we can write $B = (a, \lambda)$ for some λ.

In order to obtain the equation of the curve traced by P, we will show that x, y, λ satisfy two polynomial relations, namely

$$x^2 + y^2 = \lambda y, \quad 2x(a - x) = (2y - \lambda)(y - \lambda).$$

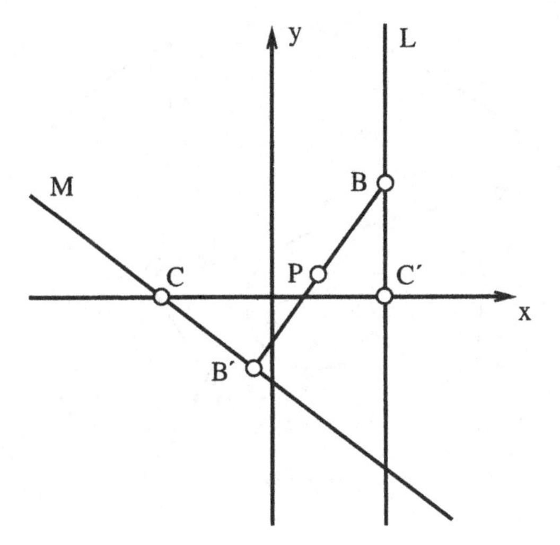

Fig. 1.10. The Newton motion

To this end write $P = (h, k)$; clearly $h = 2x - a$, $k = 2y - \lambda$. By symmetry, the distance from B to C' coincides with the distance from B' to C, producing the first displayed relation. Also, the lines BB', $B'C$ must be perpendicular; equivalently, the product of their slopes is -1; but BB' has slope $(k - \lambda)/(h - a)$, and BC' has slope $(h + a)/k$ so the condition is $h^2 + k^2 = a^2 + \lambda k$, reducing to the second displayed relation when we write $h = 2x - a$, $k = 2y - \lambda$. Eliminating λ between the two displayed relations gives $x(2a - x)y^2 = x^4$. The possibility $x = 0$ only gives rise to the case when P is the origin, so we conclude that P must lie on the cubic curve $(2a - x)y^2 = x^3$ which is indeed a cissoid of Diocles when we replace x in the formula of Example 1.8 by $-x$.

The cissoid can be physically traced by taking two sheets of wood (the fixed and moving planes) on each of which is a groove (the lines L and M) and a small fixed peg (the points C and B) designed to slide in the groove on the other sheet. The distances from peg to groove must be the same. The sheet with groove L and peg C is fixed on a work surface; the second sheet can then slide with the peg B in the groove L, and the peg C in the groove M. The cissoid is traced by the point P halfway between B and M.

More complex trajectories, of considerable geometric interest, can be generated via both constructions by allowing L, M to be more compli-

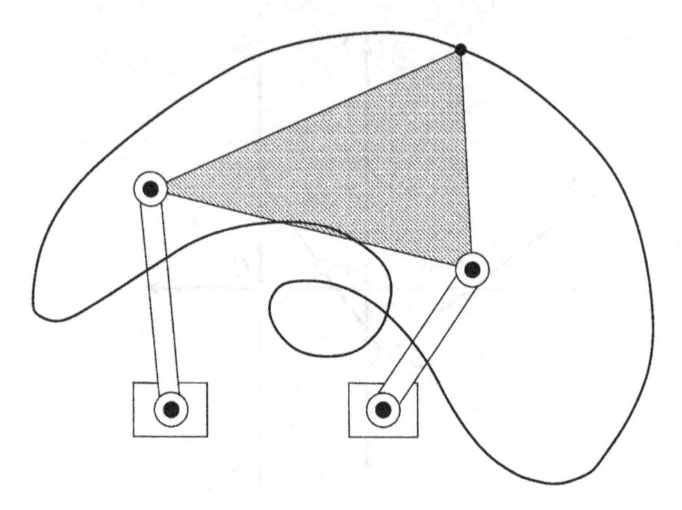

Fig. 1.11. Four bar linkage coupler curve

cated curves. In the next (and final) example B, C move along circles, and the trajectories turn out to be sextic curves.

Example 1.13 In mechanical engineering, the general *four bar linkage* is obtained by taking a quadrilateral $ABCD$ hinged at the points A, B, C, D with sides AB, BC, CD, DA of fixed lengths. Fixing one of the sides (AD for instance) we see that the point B moves round a circle centred at A, the point C moves round a circle centred at D, and the *coupler bar* BC moves with one dof. (The reader is urged to construct a linkage using cardboard pinned to a flat surface.) A tracing point P in the moving plane carried by BC then describes a trajectory, known as a *coupler curve*; given the simplicity of the construction, the complexity of this curve can be quite surprising. This example includes the Watt curve, which is a trajectory arising from a special type of four bar, with tracing point the mid-point of the coupler bar. A typical example is illustrated in Figure 1.11.

One can pursue the line of thought underlying this section much further. There is a fascinating zoo of 'mechanisms', mechanical linkages moving with one dof, which can generate trajectories of seemingly infinite complexity. Indeed there is a remarkable result (proved by a London solicitor Kempe in 1876) which says that any real algebraic curve can be mechanically generated. Very little is known about such curves, and much remains to be discovered. It is the starting point for the much wider study of spatial kinematics, likely to be of considerable practical relevance as robotics assumes an ever increasing role in our lives.

2
General Ground Fields

The purpose of this chapter is to widen the scope of our enquiry by moving from real algebraic curves defined by finite sums

$$f(x, y) = \sum_{i,j} a_{ij} x^i y^j$$

with coefficients a_{ij} in the real number field $\mathbb{R}$, to curves for which the coefficients lie in an arbitrary 'ground field' $\mathbb{K}$. There are two compelling reasons for such an extension. The first is that it will lead to a better understanding of the geometry of curves, even in the real case. And the second is that the powerful visual ideas arising from real algebraic curves can be applied to other areas of mathematics (such as number theory) to considerable effect.

2.1 Two Motivating Examples

In numerous situations one needs to count the intersections of two algebraic curves. This raises a fundamental question. How many points of intersection are there?

Example 2.1 How many times does the unit circle $f(x, y) = x^2 + y^2 - 1$ intersect the line $y = c$? For a fixed value of the constant c, the intersections are given by $x^2 = 1 - c^2$. The solutions of this equation depend critically on the value of c. For $-1 < c < 1$ there are two real solutions, for $c = \pm 1$ there is just one real solution, and for $c < -1$ and $c > 1$ there are none. (Figure 2.1.) The answer admits a more natural phrasing were we to allow the possibility of x, y assuming *complex* values. In that case we would be able to say that for $c \neq \pm 1$ there are two points of intersection, and that for $c = \pm 1$ there is just one. Better still, we could

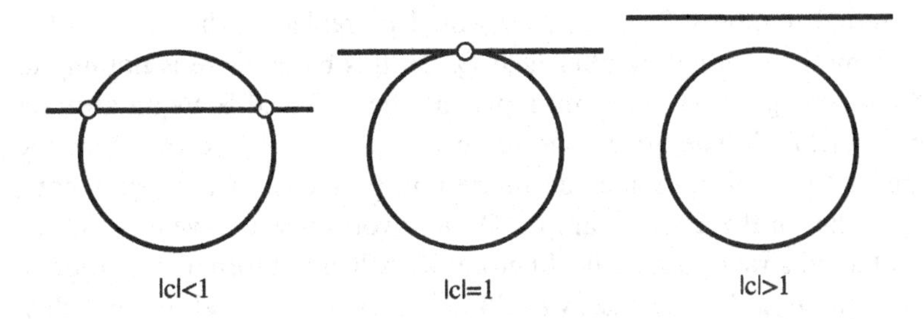

Fig. 2.1. Horizontal line meeting a circle

say that there are always two points of intersection, which coincide when $c = \pm 1$!

Such examples lead us in a significant direction. Instead of restricting our attention to points (x, y) in the real affine plane $\mathbb{R}^2$ satisfying an equation $f(x, y) = 0$ we might broaden our viewpoint to consider points (x, y) in the set $\mathbb{C}^2 = \{(x, y) : x, y \in \mathbb{C}\}$ with this property. Thus we could define a 'complex' algebraic curve in exactly the same way that we defined a 'real' algebraic curve in Section 1.1, save that the coefficients a_{ij} in the defining polynomial f would be allowed to be complex numbers. In the language of algebra, we are replacing the real number field $\mathbb{R}$ by the complex number field $\mathbb{C}$. The complex environment is both natural and fruitful, lending considerable insight into real phenomena. For instance, it will give us a simple geometric insight into the concept of a centre for a circle.

The next example opens up a rather different direction in which we might proceed, namely thinking about problems in number theory in the context of curves.

Example 2.2 A classical problem of Greek mathematics was to find all right-angled triangles with integer sides, i.e. all positive integers X, Y, Z with $X^2 + Y^2 = Z^2$. Equivalently one seeks positive *rational* solutions $x = X/Z$, $y = Y/Z$ of the equation $x^2 + y^2 = 1$. In other words, we seek all points (x, y) in the set $\mathbb{Q}^2 = \{(x, y) : x, y \in \mathbb{Q}\}$ with $x^2 + y^2 = 1$. (In Example 8.8 we will obtain a complete solution to this problem, via the idea of 'rational' parametrization.)

This example suggests it might be profitable to proceed by analogy and define a 'curve' in $\mathbb{Q}^2$ just as we did in the real case, save that the coefficients a_{ij} in the defining polynomial f would be allowed to be rational numbers. This time we would be replacing the real number field $\mathbb{R}$ by the rational number field $\mathbb{Q}$. On this basis, there is nothing to stop one going all the way and replacing the real field $\mathbb{R}$ by an abstract 'ground field' $\mathbb{K}$. The advantage of such a move is that one widens the scope of the subject to include interesting problems in number theory, such as that in the above example. Do not worry if you have not met the idea of an abstract number field before; we will give a formal definition in the next section. In any case, we will not work at this level of generality. *The convention in this book is that the underlying ground field is either the real field* $\mathbb{R}$*, or the complex field* $\mathbb{C}$. Once in a while (just for the fun of it) we may allow $\mathbb{K}$ to be some other field, but at each such juncture we will explain what the field is.

Exercises

2.1.1 A *piriform* is a quartic curve in $\mathbb{R}^2$ defined by a formula of the form $a^4y^2 = b^2x^3(2a - x)$ where a, b are positive real numbers. Determine the number of real points of intersection of the piriform with the circle $(x - a)^2 + y^2 = a^2$ in terms of a, b.

2.2 Groups, Rings and Fields

In this section we present a very brief review of the basic algebraic concepts underlying the material treated in this book. (Virtually any elementary algebra text will tell you much more.) Three algebraic structures play a significant role, namely the ideas of group, domain and field. It is natural to consider them in turn. The most basic structure is that of a 'group'. Examples of groups appear at several points in the development. For instance, the 'affine' maps' of Chapter 4 form a group under the operation of composition; and we will see in Chapter 17 that certain cubic curves have a natural group structure, the source of much interesting geometry. Here is a brief reminder of the definitions. A *group* is a set X, together with an 'operation' $\cdot$, which satisfies the following *group axioms* (G1), (G2), (G3), (G4).

- (G1) $\cdot$ is a binary operation on X, i.e. there is a rule whereby to each pair of elements x, y in X we associate another element $x \cdot y$ in X. (Closure.)
- (G2) For any elements x, y, z in X we have $x \cdot (y \cdot z) = (x \cdot y) \cdot z$. (Associative Law.)

- (G3) There exists an element 1 in X (the identity) with the property that for any x in X we have $1 \cdot x = x = x \cdot 1$. (Existence of an Identity.)
- (G4) For any element x in X there exists an element x^{-1} in X with the property that $x \cdot x^{-1} = 1 = x^{-1} \cdot x$. (Existence of Inverses.)

A binary operation $\cdot$ on X is *commutative* when $x \cdot y = y \cdot x$ for all choices of x, y in X. A group X whose binary operation $\cdot$ is commutative is said to be *Abelian*. It is a convention in group theory to use 'additive' notation when dealing with Abelian groups; thus the operation is written $+$, the identity is written 0, and the inverse of x is written $-x$. (We will use this notation when dealing with the Abelian group structure on cubic curves.) For an element x in X, and a positive integer n we define $x^n = x \cdot x \cdot \ldots \cdot x$ (n times); additively, we write $nx = x + x + \cdots + x$. An element x is said to be of *finite order* when there exists a positive integer n with $x^n = 1$ (additively, with $nx = 0$), and in that case the least such n is the *order* of x.

The most familiar examples of groups are the sets $\mathbb{Z}$ of integers, $\mathbb{Q}$ of rational numbers, $\mathbb{R}$ of real numbers, and $\mathbb{C}$ of complex numbers, all with the usual definitions of addition $+$, and with identity element 0. Likewise, the sets $\mathbb{Q}^\star$, $\mathbb{R}^\star$, $\mathbb{C}^\star$ of non-zero elements in $\mathbb{Q}$, $\mathbb{R}$, $\mathbb{C}$ are Abelian groups under the usual definitions of multiplication $\cdot$, with identity element 1. A more exotic example is the group $\mathbb{Z}_p$ of residue classes of integers modulo p, obtained as follows.

Example 2.3 Let p be a fixed positive integer, and let $\mathbb{Z}$ be the set of integers. Recall that two integers a, b are said to be *congruent* modulo p, and we write $a \equiv b \bmod p$, when $a - b$ is divisible by p. This defines an equivalence relation partitioning $\mathbb{Z}$ into disjoint equivalence classes, known as *residue classes* of integers. The residue class $\overline{a}$ containing a comprises all integers congruent to a mod p: thus there are p residue classes $\overline{0}$, $\overline{1}$, ... , $\overline{(p-1)}$ modulo p, forming the set $\mathbb{Z}_p$ of all residue classes modulo p. Addition of residue classes is defined by $\overline{a} + \overline{b} = \overline{a+b}$. It should be clear from the definitions that $\mathbb{Z}_p$ is an Abelian group under the operation $+$ with identity element $\overline{0}$, and with inverses defined by $-\overline{a} = \overline{(-a)}$.

Example 2.4 Write $\mathbb{Z}_p^\star$ for the set of non-zero elements in $\mathbb{Z}_p$. Here we need to proceed more carefully. Multiplication of residue classes is defined by $\overline{a} \cdot \overline{b} = \overline{a \cdot b}$. We claim that *provided p is prime, $\mathbb{Z}_p^\star$ is an Abelian group* under multiplication with identity element $\overline{1}$. It will not be too confusing simply to write a for the residue class $\overline{a}$. Note first that if $a \cdot b = 0$ then either $a = 0$ or $b = 0$, for if $a \cdot b = 0$ then p is a factor of ab, so p is a factor of a or a factor of b (that is the point where we use the assumption that p is prime) and hence $a = 0$ or $b = 0$. It follows that $\mathbb{Z}_p^*$ is closed under multiplication. It remains to establish the existence of inverses. We have to show that if $a \neq 0$ then there exists an element b such that $a \cdot b = 1$. To this end, note that the $(p-1)$ elements $a \cdot 1, \ldots, a \cdot (p-1)$ are non-zero and distinct: for if we had an equality $a \cdot j = a \cdot k$ with $j \neq k$ then $a \cdot (j-k) = 0$, and hence $j = k$, a contradiction. However, there are exactly $(p-1)$ elements in $\mathbb{Z}_p^*$, so they must be $a \cdot 1, \ldots, a \cdot (p-1)$. In particular, the element 1 must be in the sequence, i.e. $a \cdot b = 1$ for some $b \neq 0$.

For the purposes of this book a *commutative ring* will mean a set $\mathbf{R}$ with distinct elements 0, 1 and commutative operations $+$ (*addition*) and $\cdot$ (*multiplication*) satisfying the following 'axioms', of which (3) is the Distributive Law.

(1) $\mathbf{R}$ is a group under the operation $+$ with identity 0.
(2) Multiplication $\cdot$ is associative with identity 1.
(3) $x \cdot (y+z) = x \cdot y + x \cdot z$ for all x, y, z in $\mathbf{R}$.

A commutative ring $\mathbf{R}$ is a *domain* when in addition we have no *divisors of zero*, i.e. for any choices of a, b the relation $ab = 0$ implies that $a = 0$ or $b = 0$. Cancellation is possible in domains, meaning that if $ab = ac$ and $a \neq 0$ then $b = c$; just observe that $ab = ac$ is equivalent to $a(b-c) = 0$. An element u in $\mathbf{R}$ is a *unit* when it has a multiplicative inverse, i.e. an element u^{-1} (necessarily unique) with $u \cdot u^{-1} = u^{-1} \cdot u = 1$. A commutative ring $\mathbf{R}$ is a *field* when every element $a \neq 0$ has a multiplicative inverse; thus a field is automatically a domain.

Important examples of domains for us are $\mathbb{Z}$, $\mathbb{Q}$, $\mathbb{R}$ and $\mathbb{C}$, all with the usual definitions of addition and multiplication; of these $\mathbb{Q}$, $\mathbb{R}$ and $\mathbb{C}$ are fields. But for fun purposes we will add the set $\mathbb{Z}_p$ of residue classes of integers modulo a prime number p. The first case of interest is then the field $\mathbb{Z}_3$, comprising the residue classes 0, 1, 2 with addition and multiplication tables as below.

+	0	1	2
0	0	1	2
1	1	2	0
2	2	0	1

·	0	1	2
0	0	0	0
1	0	1	2
2	0	2	1

The number field $\mathbb{Z}_p$ is distinguished from the number fields $\mathbb{Q}$, $\mathbb{R}$ and $\mathbb{C}$ by the fact that it has 'finite characteristic'. Generally, a domain $\mathbb{D}$ has *finite characteristic* when there exists a positive integer p for which $0 = 1 + 1 + \cdots + 1$ (p times); and in that case the least positive integer p for which this holds is the *characteristic* of $\mathbb{D}$. If no such integer exists then $\mathbb{D}$ is said to have *characteristic zero*. Thus $\mathbb{Z}_p$ has characteristic p, whilst $\mathbb{Q}$, $\mathbb{R}$ and $\mathbb{C}$ have characteristic zero.

Example 2.5 The process whereby the field of rationals $\mathbb{Q}$ is obtained from the domain $\mathbb{Z}$ of integers by forming 'fractions' can be extended to any domain $\mathbb{D}$. Consider the set of pairs (a, b) with $a, b \in \mathbb{D}$ and $b \neq 0$. On this set we introduce an equivalence relation $\sim$ by agreeing that $(a, b) \sim (c, d)$ if and only if $ad = bc$. Equivalence classes are called *fractions*, and we denote the class containing the pair (a, b) by a/b. Addition $+$ and multiplication $\cdot$ of fractions are then defined by

$$\frac{a}{b} + \frac{c}{d} = \frac{ad + bc}{bd}, \qquad \frac{a}{b} \cdot \frac{c}{d} = \frac{ac}{bd}.$$

With these definitions it is easily verified that the fractions form a field, the *field of fractions* over the domain $\mathbb{D}$. And by identifying the element $a \in \mathbb{D}$ with the fraction $a/1$, we identify $\mathbb{D}$ with a subset of the field of fractions.

2.3 General Affine Planes and Curves

At this point it may be helpful to embark on a little harmless abstraction, just to clarify the foundations of what we are doing. Consider a set X, whose elements we will call *points*, together with a collection of non-empty subsets of X called *lines*. We say that a point p *lies on* a line l, or that l *passes through* p, when p is an element of the subset l. We refer to X as a *plane* when points and lines satisfy the following axioms.

- There are at least three points which do not lie on any one line.
- Through any two points p, q there passes exactly one line l.

Let X be a plane, let l be a line in X, and let p be a point in X. Two lines are said to be *parallel* when they are disjoint subsets of X. In principle we have three posssibilities.

(i) There are no lines through p parallel to l.
(ii) There is exactly one line through p parallel to l.
(iii) There is more than one line through p parallel to l.

The most familiar situation is when X is an *affine* plane, meaning that (ii) holds for all choices of line l, and point p not on l. Later in this book we will meet *projective* planes, where (i) holds for all choices of line l, and point p not on l: in other words, a projective plane is one in which any two lines intersect in at least one point. The next example constructs a wide range of affine planes, sufficient for all our purposes. (The reader is warned that not all affine planes arise in this way.)

Example 2.6 Let $\mathbb{K}$ be a field, and let $\mathbb{K}^2 = \{(x, y) : x, y \in \mathbb{K}\}$. By a *line* in $\mathbb{K}^2$ we mean a subset of the form $\{(x, y) \in \mathbb{K}^2 : ax + by + c = 0\}$ where a, b, c are elements of $\mathbb{K}$ with at least one of $a, b \neq 0$. With this definition $\mathbb{K}^2$ is a plane. Clearly, there is no line through the three points $(0,0)$, $(1,0)$, $(0,1)$: and the argument in Example 1.1, establishing that there is a unique line through any two distinct points in $\mathbb{R}^2$, works over any field. Indeed $\mathbb{K}^2$ is an *affine* plane; the argument in Example 1.3 also works over any field. With this concept of 'line', we call $\mathbb{K}^2$ the affine plane over the *ground field* $\mathbb{K}$. The plane $\mathbb{R}^2$ is known as the *real* affine plane, and $\mathbb{C}^2$ as the *complex* affine plane.

The main point to make is that it is only the affine plane structure of $\mathbb{K}^2$ which plays a role in this book. For instance, the metric structure on the real affine plane $\mathbb{R}^2$ given by the concepts of distance and angle is irrelevant. (Very occasionally we use the metric structure on $\mathbb{R}^2$, but only as a convenient device to define interesting examples of real algebraic curves.) Here is a rather odd example of an affine plane. Although it is really a 'fun' example, we will meet it later in an unexpected context, namely the geometry of cubic curves.

Example 2.7 Consider the plane $\mathbb{K}^2$, where $\mathbb{K}$ is the finite field $\mathbb{Z}_3$ with elements 0, 1, 2. There are nine points in $\mathbb{K}^2$, namely $(0,0)$, $(0,1)$, $(0,2)$, $(1,0)$, $(1,1)$, $(1,2)$, $(2,0)$, $(2,1)$ and $(2,2)$. It is convenient to picture them as the nine points in $\mathbb{R}^2$ with the same coordinates. (Figure 2.2.) There are twelve lines: three 'vertical' lines $x = 0$, $x = 1$, $x = 2$, three 'horizontal' lines $y = 0$, $y = 1$, $y = 2$, two 'diagonal' lines $y = x$, $y = -x + 2$, and four more $y = x + 1$, $y = -x$, $y = x - 1$, $y = -x + 1$. Thus for instance the zero set of the line $y = x + 1$ comprises the points $(0,1)$, $(1,2)$ and $(2,0)$.

We can define 'curves' in general affine planes $\mathbb{K}^2$ just as we did in the real affine plane $\mathbb{R}^2$. A polynomial $f(x, y)$ in the variables x, y over the

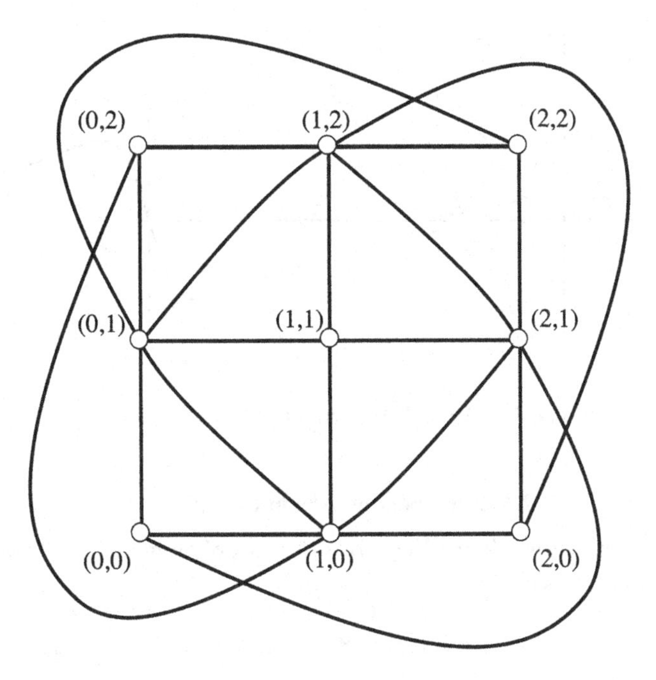

Fig. 2.2. The affine plane over $\mathbb{Z}_3$

field $\mathbb{K}$ is defined to be a formula of the shape

$$f(x, y) = \sum_{i,j} a_{ij} x^i y^j$$

where the sum is finite and the coefficients a_{ij} lie in $\mathbb{K}$. An *algebraic curve* over $\mathbb{K}$ is a non-zero polynomial $f(x, y)$ over $\mathbb{K}$, *up to multiplication by a non-zero scalar*. Henceforth, we will abbreviate the term 'algebraic curve' to 'curve'. The *degree* of a curve is the common degree of its defining polynomials. Curves of degree 1, 2, 3, 4, ... are called *lines*, *conics*, *cubics*, *quartics*, ... Roughly speaking, the higher the degree of a curve the more complicated it may become, so there is a lot to be said for understanding curves of very low degree in some detail. That is part of our underlying philosophy. Thus in Chapter 5 we will present a thorough discussion of conics. And cubics receive a lot of attention, culminating in the beautiful connexion between cubic curves and groups in Chapter 17.

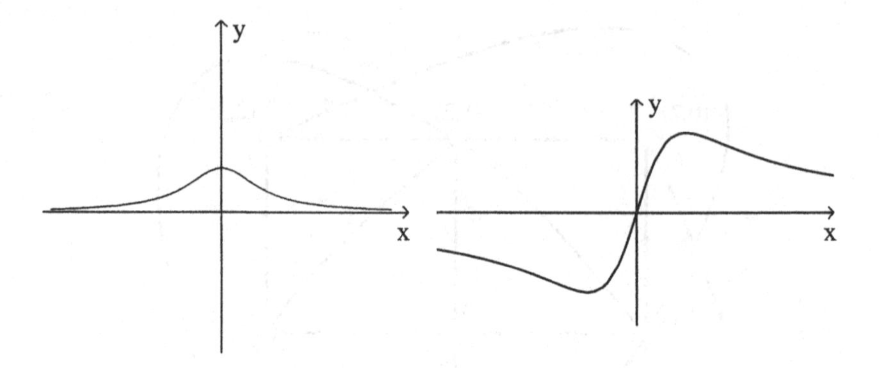

Fig. 2.3. Agnesi's versiera and the Serpentine

2.4 Zero Sets of Algebraic Curves

As in Section 1.1, we define the *zero set* of a polynomial $f(x,y)$ over the field $\mathbb{K}$ to be the set

$$V_f = \{(x,y) \in \mathbb{K}^2 : f(x,y) = 0\}.$$

The zero set is unchanged when we multiply f by a non-zero scalar, so is defined not just for polynomials, but for the associated curves. Consider first the case of the real field $\mathbb{R}$. In Chapter 1, we sketched a number of zero sets of real algebraic curves. Generally speaking, it can be quite difficult to do this by hand, indeed there was a time (not so long ago) when mathematical texts were devoted to this art. However, life has moved on; in principle zero sets can be traced by computer within a fraction of a second. All the same, there is much to be said for being able to sketch simple examples. Curves which can be put in the form $y = \phi(x)$ for some (not necessarily polynomial) function $\phi(x)$ are an obvious starting point.

Example 2.8 The zero set of a curve $x^2y = 4a^2(2a - y)$ with $a > 0$ (this is Agnesi's versiera of Exercise 1.2.6) is rather easy to sketch, because it is the graph of the function $y = 8a^3/(x^2 + 4a^2)$ depicted in Figure 2.3. Likewise the zero set of the *serpentine* $x^2y + a^2y - b^2x$ with $a, b > 0$ is the graph of the function $y = b^2x/(x^2 + a^2)$.

The point of the next example is that it provides you with another useful class of curves to which elementary sketching techniques apply.

Example 2.9 Consider curves of the form $y^2 = \phi(x)$, with $\phi(x)$ a (not necessarily polynomial) function. The method is first to sketch the graph of $\phi(x)$, and then to 'take square roots'. A concrete example is provided by the family of curves $y^2 = x^3 - 3x + \lambda$, where λ is a real parameter. Here $\phi(x) = x^3 - 3x + \lambda$. Think of this as the function $y = x^3 - 3x$ translated upwards (or downwards) by λ. The polynomial $y = x^3 - 3x$ is a cubic with three real roots having a local maximum at $x = -1$ with value $y = 2$, and a local minimum at $x = 1$ with value $y = -2$. Figure 2.4 shows how the graph of $\phi(x)$ varies as λ varies. In each picture we now 'take square roots', to obtain the series of pictures of the curve $y^2 = \phi(x)$ as λ varies. We will return to this class of examples later.

One of the psychological difficulties of working over general ground fields (as opposed to the real field) is that there may be no way of visualizing the zero set. However, one can soon adjust to that. The main point to make is that the analogy with the real case is a powerful tool for understanding a diversity of mathematical problems.

Example 2.10 We will determine the zero set of the conic $x^2 - y^2 = 1$, over the field $\mathbb{Z}_3$. There are only three possible values for y, namely $y = 0$, $y = 1$, $y = 2$. When $y = 0$ we have $x^2 = 1$, and hence $x = 1$ or $x = 2$, and when $y = 1$, $y = 2$ we have $x^2 = 2$, with no solutions. Thus the zero set comprises the points $(1, 0)$, $(2, 0)$.

In this book we adopt a rather geometric viewpoint, by which we mean that our intuitions are based on the visual properties of zero sets for real algebraic curves. However, we have to bear in mind that the underlying object is the polynomial since, although the curve determines the zero set, the converse may not hold. Here is a salutary example.

Example 2.11 The conics $x^2 + y^2 + 1$, $x^2 + y^2 + 2$ in $\mathbb{R}^2$ are distinct (clearly, neither is a scalar multiple of the other) but have the *same* zero set, namely the empty set. The trouble here is that we are working over the real field. Over the complex field the zero sets are distinct; for instance, the point $(i, 0)$ lies on the first curve but not on the second. A different phenomenon is illustrated by the cubic curves x^2y, xy^2 which are distinct, but have identical zero sets; this time the trouble is that the equations factorize.

In Chapter 14, we will return to the question of the relation between curves and their zero sets, and show that in the complex case the zero set determines the curve, provided the equation does not factorize. Until then

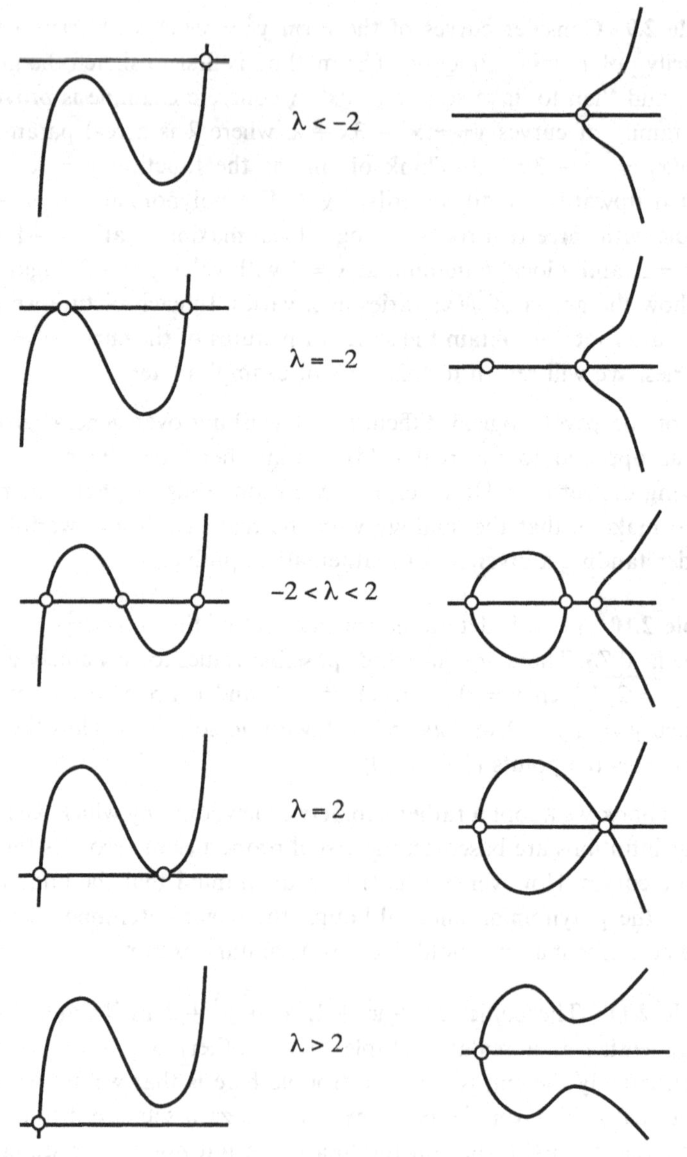

Fig. 2.4. The curves $y^2 = x^3 - 3x + \lambda$

the reader is advised to heed the warning given in Chapter 1, namely that one needs to maintain a very clear distinction between the two concepts. Example 2.11 makes the point that real algebraic curves may have empty zero sets. By contrast, zero sets of complex algebraic curves always have lots of points!

Lemma 2.1 *The zero set of any curve f of degree $d \geq 1$ in $\mathbb{C}^2$ is infinite.*

Proof The idea is to show that infinitely many lines $y =$ constant meet the curve at least once. The proof depends on the fact that any polynomial $\phi(x) = \sum_{i=0}^{d} a_i x^i$ of degree $d \geq 1$ has at least one *zero*, i.e. a complex number x which satisfies the equation $\phi(x) = 0$. (That is the Fundamental Theorem of Algebra: see Lemma 3.7.) Write $f(x, y) = \sum_{i=0}^{d} a_i(y)x^i$ where the coefficients $a_i(y)$ are polynomials in y. We can suppose at least one polynomial $a_i(y)$ with $i \geq 1$ is non-zero; otherwise $f(x, y) = a_0(y)$, and the zero set is a union of lines $y =$ constant, so infinite. For fixed y the equation $f(x, y) = 0$ is polynomial in x with complex coefficients, and has a zero x, provided it has degree ≥ 1. That only fails when $0 = a_1(y) = \ldots = a_d(y)$. But a non-zero polynomial has at most finitely many zeros, so there are only finitely many 'exceptional' values y for which these conditions can hold. There are therefore infinitely many y for which the conditions fail, and hence for which the equation $f(x, y) = 0$ has at least one solution (x, y). □

Example 2.12 Consider the conic $f(x, y) = xy - 1$ in $\mathbb{C}^2$. With the notation of the above lemma $a_0(y) = -1$, $a_1(y) = y$. The only 'exceptional' value of y is given by $0 = a_1(y) = y$: indeed, the line $y = 0$ does not meet the zero set, whereas any line $y = c$ with $c \neq 0$ does.

It is interesting to take the very simplest algebraic curves in $\mathbb{C}^2$ and determine their zero sets explicitly.

Example 2.13 The zero set of a line $ax + by + c$ in $\mathbb{C}^2$ is a real plane! Identifying a point $x = x_1 + ix_2$ in $\mathbb{C}$ with a point (x_1, x_2) in $\mathbb{R}^2$, and likewise a point $y = y_1 + iy_2$ in $\mathbb{C}$ with a point (y_1, y_2) in $\mathbb{R}^2$, we identify a point (x, y) in $\mathbb{C}^2$ with a point (x_1, x_2, y_1, y_2) in $\mathbb{R}^4$. The single complex linear equation $ax + by + c = 0$ then represents two real linear equations in x_1, x_2, y_1, y_2; indeed, if we write $a = a_1 + ia_2$, $b = b_1 + ib_2$, $c = c_1 + ic_2$ and equate real and imaginary parts we obtain

$$\begin{aligned} a_1x_1 - a_2x_2 + b_1y_1 - b_2y_2 &= -c_1 \\ a_2x_1 + a_1x_2 + b_2y_1 + b_1y_2 &= -c_2. \end{aligned}$$

representing *two* linear equations in the *four* unknowns x_1, x_2, y_1, y_2. The matrix of this system is

$$\begin{pmatrix} a_1 & -a_2 & b_1 & -b_2 \\ a_2 & a_1 & b_2 & b_1 \end{pmatrix}.$$

Suppose the matrix of coefficients has rank < 2. Then (by linear algebra) all the 2×2 minors vanish. In particular, $a_1^2 + a_2^2 = 0$ and $b_1^2 + b_2^2 = 0$, so $a_1 = a_2 = 0$, $b_1 = b_2 = 0$. But that means $a = b = 0$, which is impossible, as one of a, b must be non-zero. It follows that the matrix has rank two, and (by linear algebra again) the solution space of the equations is an affine plane in $\mathbb{R}^4$.

This is the first instance of a general principle, namely that *curves in $\mathbb{C}^2$ can be considered as real surfaces in $\mathbb{R}^4$*; in the single complex relation $f(x, y) = 0$ defining an algebraic curve in $\mathbb{C}^2$ we can take real and imaginary parts to produce two polynomials in the variables x_1, x_2, y_1, y_2 defining (in principle) a surface in $\mathbb{R}^4$.

Exercises

2.4.1 Use the sketching technique of Example 2.9 to sketch the zero sets of the following real algebraic curves.

(i)	$f = y^2 - x^3$	(v)	$f = y^2 - x^2 + x^4$
(ii)	$f = y^2 + x^2 - x^3$	(vi)	$f = y^2 - x^3 + x^4$
(iii)	$f = y^2 - x^2 - x^3$	(vii)	$f = y^2 + x^3 - x^4$
(iv)	$f = y^2 + x^2 - x^4$	(viii)	$f = x^2 + y^2 - x^2y^2$.

2.4.2 Sketch the zero set of the real algebraic curve given by the formula $f = x(x^2 + y^2) - (1 - x)^3$.

2.4.3 Let $p(x)$ be a polynomial over the field $\mathbb{K}$. Show that the graph of $p(x)$ in $\mathbb{K}^2$ is the zero set of an algebraic curve.

2.4.4 Find the zero sets of the conics $f_0 = x^2 + y^2$, $f_1 = x^2 + y^2 + 1$ and $f_2 = x^2 + y^2 + 2$ over the field $\mathbb{Z}_3$.

2.4.5 Find the zero set of the line $2x + 3y + 1$ over the field $\mathbb{Z}_5$. Express the line in the form $y = \alpha x + \beta$, and hence check your answer.

2.4.6 Let f be a curve of *odd* degree d in $\mathbb{R}^2$ with $f(0,0) \neq 0$. Show that there are at most finitely many real numbers λ for which f fails to intersect the line $y = \lambda x$, and deduce that the zero set of f is infinite. (You may assume that a real polynomial of odd degree has at least one zero.)

3

Polynomial Algebra

In this chapter, we lay some algebraic foundations for a systematic study of curves, specifically the question of factorizing polynomials in one or more variables. Later, we will return to more special topics in algebra, but till then this material will serve all our needs. We start by discussing the abstract concept of factorization in an arbitrary domain. In the succeeding section, we recall the elementary algebra associated to polynomials in a single variable; that provides a model for the algebra of polynomials in several variables, and is anyway of particular relevance to the geometry of curves. We devote a separate section to the special polynomials known as 'forms', crucial to understanding 'tangents' and 'projective' curves. Finally, we explain how some basic calculus ideas can be introduced for polynomials over arbitrary domains.

3.1 Factorization in Domains

Much of this book will centre around the concept of 'factorizing' polynomials. The domain $\mathbb{Z}$ of integers provides the model for the ideas, but we phrase the definitions for any domain $\mathbb{D}$. Given elements $a, b \in \mathbb{D}$ we say that a is a *factor* of b, or that b is *divisible* by a, written $a|b$, when there exists an element $c \in \mathbb{D}$ with $b = ac$. (Note that according to this definition the zero element 0 is divisible by any element in $\mathbb{D}$.) A factor is said to be *trivial* when it is a unit, otherwise it is *proper*. The *associates* of an element a are the elements of the form au with u a unit. (Thus in $\mathbb{Z}$ the units are ± 1.) There are two key ideas in discussing factorization:

- An element $p \in \mathbb{D}$ is *irreducible* when $p \neq 0$, p is not a unit, and $p = ab$ means that one of a, b is a unit.

- An element p is *prime* when $p \neq 0$, p is not a unit, and $p|ab$ means that either $p|a$ or $p|b$.

A prime element is necessarily irreducible, though in general irreducibles may fail to be prime. (Exercise 3.1.1.) A *unique factorization domain* (UFD) is a domain $\mathbb{D}$ with the following properties:

- Any element $a \in \mathbb{D}$ can be written as a finite product $a = a_1 \ldots a_p$ where the factors $a_1, \ldots, a_p$ are irreducible.
- The factorization into irreducibles is unique, in the sense that if $a = a_1 \ldots a_p$, $a = a'_1 \ldots a'_q$ are two factorizations into finitely many irreducibles, then $p = q$, and each a_i is associate to exactly one a'_j.

In a unique factorization domain the concepts of prime and irreducible coincide. (Exercise 3.1.1.) The most familiar example is the domain $\mathbb{Z}$ of integers where these concepts yield the 'prime numbers' of elementary number theory. One of the central facts we will use in this book is that polynomial domains (to be introduced in the succeeding sections) are likewise unique factorization domains; the next two abstract lemmas will pinpoint the key ingredient in this result. The first provides a sufficient condition for the *existence* of factorizations into finitely many *irreducible* elements. First a definition. A *degree* function on a domain $\mathbb{D}$ is a function d on $\mathbb{D}$ with non-negative integer values, such that $d(a) < d(ab)$, when b is not a unit. The model for a degree function is the absolute value function on the domain $\mathbb{Z}$ of integers.

Lemma 3.1 *Let $\mathbb{D}$ be a domain on which there is a degree function. Then any element f can be written as a product $f = f_1 \ldots f_s$ of irreducibles.*

Proof If f is irreducible, we are done. Otherwise write $f = gh$ where g, h are proper factors, so $d(g) < d(gh) = d(f)$, $d(h) < d(gh) = d(f)$. Now repeat this process to g and to h; if either fails to be irreducible we factor it into proper factors on which d takes strictly smaller values. The process can be repeated at most finitely many times, since at each stage the degree function drops by at least one. The result follows. □

Here is an equally simple result concerning the *uniqueness* of polynomial factorizations into *prime* elements.

Lemma 3.2 *Let $\mathbb{D}$ be a domain, let $f \neq 0$ be an element which is not a unit, and let $f = f_1 \ldots f_p = g_1 \ldots g_q$ be factorizations of f into primes.*

Then $p = q$, each of the f's is an associate of one of the g's, and conversely each of the g's is an associate of one of the f's.

Proof Since f_1 is prime and divides $g_1 \dots g_q$ it must be a factor of some g_j. We can suppose the g's numbered so that $j = 1$. Since g_1 is prime that means f_1, g_1 are associates, and $g_1 = uf_1$ for some unit $u \neq 0$. Substituting for g_1 in $f_1 \dots f_p = g_1 \dots g_q$, and cancelling f_1, we obtain a relation $f_2 \dots f_p = ug_2 \dots g_q$. Repeating this argument p times we obtain a relation $1 = g_{q-p} \dots g_q$. That implies that $p \leq q$ since the g's are not units. Likewise we establish that $q \leq p$ and hence $p = q$. In the process we have established that each f is an associate of some g, and conversely. □

Combining these two results, we obtain the following sufficient condition for a domain to have unique factorization.

Lemma 3.3 *Let $\mathbb{D}$ be a domain. Assume that $\mathbb{D}$ has a degree function, and that every irreducible is prime. Then $\mathbb{D}$ is a unique factorization domain. In particular, any element $f \neq 0$ can be written uniquely (up to order and units) in the form $f = cf_1^{r_1} \dots f_s^{r_s}$ where c is a unit, where $f_1, \dots ,f_s$ are irreducible, and for $i \neq j$ no f_i is a factor of f_j.*

Exercises

3.1.1 Let $\mathbb{D}$ be a domain. Show that a prime element in $\mathbb{D}$ is necessarily irreducible. And show that if $\mathbb{D}$ is a unique factorization domain then conversely any irreducible element is prime.

3.2 Polynomials in One Variable

It is worth recalling the generalities of factorization for polynomials in a single variable x over a domain $\mathbb{D}$. The case when $\mathbb{D}$ is the domain of real numbers is probably familiar ground for the reader. All we are doing in this section is simultaneously to recall school algebra, and phrase it in more general terms, for reasons which will appear later when we progress to polynomials in several variables. A *monomial* in x is defined to be a power x^k: and a *polynomial* in x over the domain $\mathbb{D}$ is defined to be a linear combination of monomials in x with coefficients in $\mathbb{D}$, so is given by a formula $f(x) = a_0 + a_1x + \cdots + a_dx^d$, where the coefficients $a_i \in \mathbb{D}$. The *degree* of a monomial x^k is defined to be k, and that of a polynomial $f(x)$ is the maximal degree of the monomials x^k for which the associated coefficient $a_k \neq 0$. Polynomials of the form $f(x) = a$ with $a \in \mathbb{D}$ are *constant* with *value a*. We can add and multiply polynomials in an 'obvious' way. Given two polynomials

$$f(x) = a_0 + a_1 x + a_2 x^2 + \cdots, \quad g(x) = b_0 + b_1 x + b_2 x^2 + \cdots$$

we define their *sum* $f + g$, and their *product* $f.g$ to be the polynomials given by

$$\begin{cases} (f+g)(x) &= c_0 + c_1 x + c_2 x^2 + \cdots \\ (f.g)(x) &= d_0 + d_1 x + d_2 x^2 + \cdots \end{cases}$$

where $c_k = a_k + b_k$ and $d_l = a_0 b_l + \cdots + a_l b_0$. With these definitions of addition and multiplication, it is easily checked that the polynomials in x over a domain $\mathbb{D}$ form another domain $\mathbb{D}[x]$. The zero element is the constant polynomial with value 0, and the identity element is the constant polynomial with value 1. The units in $\mathbb{D}[x]$ are the constant polynomials with value a unit in $\mathbb{D}$. In the case when $\mathbb{D}$ is a *field*, the units are the constant polynomials with non-zero values. It is easily checked that the degree of a polynomial is *additive*, in the sense that for polynomials $f_1, \ldots, f_p$ the degree of the product $f = f_1 \ldots f_p$ is given by

$$\deg f = \deg f_1 + \cdots + \deg f_p.$$

Note that the concept of 'irreducibility' depends on the domain $\mathbb{D}$; for instance, the polynomial $f(x) = x^2 + 1$ is irreducible over the real field, but reducible over the complex field since $x^2 + 1 = (x - i)(x + i)$. The starting point for studying polynomials in one variable is the following result from elementary algebra.

Lemma 3.4 *Let $f(x) = \sum_{i=0}^{d} a_i x^i$ be a non-constant polynomial of degree d over a domain $\mathbb{D}$, and let c be a scalar. There exists a polynomial $q(x)$ (the quotient) of degree $(d-1)$, and a scalar r (the remainder) with $f(x) = (x-c)q(x) + r$.* (The Remainder Theorem.)

Proof It suffices to show that each monomial x^i can be written in the form $x^i = (x-c)q_i(x) + r_i$ with $q(x)$ a polynomial of degree $(i-1)$, and r_i a scalar, since then

$$f(x) = \sum a_i x^i = \sum a_i\{(x-c)q_i(x) + r_i\} = (x-c)q(x) + r$$

where $q(x) = \sum a_i q_i(x)$ has degree $(d-1)$, and $r = \sum a_i r_i$ is a scalar. It remains to observe that we can take $q_i(x) = x^{i-1} + x^{i-2}c + \cdots + c^{i-1}$ and $r_i = c^i$. □

We will use the Remainder Theorem in the special case when c is a *zero* of $f(x)$, meaning that $f(c) = 0$; alternatively, we will say that c is a *root* of the equation $f(x) = 0$.

Lemma 3.5 *Let c be a zero of the polynomial $f(x)$ over the domain $\mathbb{D}$. Then $(x - c)$ is a factor of $f(x)$, i.e. we can write $f(x) = (x - c)q(x)$ for some polynomial $q(x)$ over $\mathbb{D}$.* (The Factor Theorem.)

Proof Taking $x = c$ in the Remainder Theorem we see that the remainder $r = f(c) = 0$, since c is assumed to be a zero. Thus $f(x) = (x - c)q(x)$. □

Lemma 3.6 *Any non-constant polynomial $f(x)$ of degree d over a domain $\mathbb{D}$ has at most d distinct zeros.*

Proof Let $c_1, \ldots, c_d$ be distinct zeros of $f(x)$. Then the Factor Theorem tells us that $f(x) = (x - c_1)g(x)$ for some polynomial $g(x)$ of degree $(d - 1)$. Since c_2 is a zero of $f(x)$ we have $0 = f(c_2) = (c_2 - c_1)g(c_2)$, and hence c_2 is a zero of $g(x)$. (This uses the fact that there are no divisors of zero in a domain $\mathbb{D}$.) Using the Factor Theorem again we get $f(x) = (x - c_1)(x - c_2)h(x)$ for some polynomial $h(x)$ of degree $(d - 2)$. Continuing in this way we eventually obtain $f(x) = c(x - c_1) \cdots (x - c_d)$ for some scalar $c \neq 0$. It is then immediate that any zero of $f(x)$ must be one of $c_1, \ldots, c_d$. □

Exercise 3.2.2 shows that Lemma 3.6 fails for a polynomial over a general commutative ring. We can say more about factorization of polynomials in one variable for special types of domain. A classic result from complex analysis is

Lemma 3.7 *Every non-constant polynomial $f(x)$ over the complex field has at least one zero.* (The Fundamental Theorem of Algebra.)

Generally speaking, a field $\mathbb{K}$ is said to be *algebraically closed* when any non-constant polynomial $f(x)$ over $\mathbb{K}$ has at least one zero in $\mathbb{K}$. (So the above result can be paraphrased by saying that the complex field $\mathbb{C}$ is algebraically closed.) For such a field we obtain a factorization $f(x) = c(x - \beta_1) \cdots (x - \beta_d)$ where the zeros $\beta_1, \ldots, \beta_d$ are not necessarily distinct, and $c \neq 0$. Let $\gamma_1, \ldots, \gamma_s$ be the *distinct* zeros of f occurring respectively $r_1, \ldots, r_s$ times. ($r_1, \ldots, r_s$ are called the *multiplicities* of the zeros.) Then we arrive at the following result, which is a special case of the Unique Factorization Theorem for general polynomials, to be discussed in the next section.

Lemma 3.8 *Any non-constant polynomial $f(x)$ over the complex field can be written uniquely (up to order) in the following form, for some complex number $c \neq 0$, and distinct complex numbers $\gamma_1, \ldots, \gamma_s$.*

$$f(x) = c(x-\gamma_1)^{r_1} \ldots (x-\gamma_s)^{r_s}.$$

Example 3.1 Let $n \geq 1$ be an integer. Then the polynomial $f(x) = x^n - 1$ has exactly n distinct zeros in $\mathbb{C}$, namely $1, \omega, \omega^2, \ldots, \omega^{n-1}$ where $\omega = e^{2\pi i/n}$. We refer to these zeros as the *complex nth roots of unity*.

Exercises

3.2.1 Show that the units in the domain $\mathbb{D}[x]$ are precisely the constant polynomials with value a unit in $\mathbb{D}$.

3.2.2 Verify that the quadratic polynomial $f(x) = x^2$ over $\mathbb{Z}_9$ has three distinct zeros; also, verify that the quadratic polynomial $g(x) = 2x(x-1)$ over $\mathbb{Z}_4$ has four distinct zeros.

3.2.3 Show that an algebraically closed field $\mathbb{K}$ has infinitely many elements. (Hint: assume $\mathbb{K}$ is finite, with elements $k_1, \ldots, k_s$, and consider the polynomial $f(x) = 1 + (x-k_1)\ldots(x-k_s)$.) Deduce that the zero set of an algebraic curve in $\mathbb{K}^2$ is infinite. (This generalizes Lemma 2.1.)

3.3 Polynomials in Several Variables

Formally, a *polynomial* $f(x_1,\ldots,x_n)$ over the domain $\mathbb{D}$ in the variables $x_1, \ldots, x_n$ is an expression given by a formula of the form

$$f(x_1,\ldots,x_n) = \sum_{i_1,\ldots,i_n} a_{i_1\ldots i_n} x_1^{i_1} \ldots x_n^{i_n}$$

where the sum is finite, and the coefficients $a_{i_1\ldots i_n}$ lie in $\mathbb{D}$. Polynomials of the form $f(x_1,\ldots,x_n) = a$ with $a \in \mathbb{D}$ are said to be *constant* with value a. The polynomials in $x_1, \ldots, x_n$ form a domain $\mathbb{D}[x_1,\ldots,x_n]$ under the 'obvious' definitions of addition and multiplication, with zero element the constant polynomial taking the value 0, and identity element the constant polynomial taking the value 1. The units in $\mathbb{D}[x_1,\ldots,x_n]$ are the constant polynomials whose values are units in $\mathbb{D}$. In the case when $\mathbb{D}$ is a *field* the units are the constant polynomials with non-zero values, and the associates of a polynomial are its non-zero scalar multiples.

Example 3.2 Recall from Example 2.5 that to any given domain is associated a 'field of fractions'. An example is represented by the domain $\mathbb{D}[x_1,\ldots,x_n]$ of polynomials in the variables $x_1, \ldots, x_n$ over a domain $\mathbb{D}$. In that case the field of fractions is denoted $\mathbb{D}(x_1,\ldots,x_n)$, and comprises fractions f/g, where f, g are polynomials in $x_1, \ldots, x_n$ with $g \neq 0$. Such fractions are known as *rational* functions in $x_1, \ldots, x_n$. The simplest case is provided by rational functions $f(x)/g(x)$ in a single variable, with $f(x)$, $g(x)$ polynomials in x. Rational functions in a single variable will play a significant role in discussing the concept of a 'rational' curve in Chapter 8.

Note one generality about polynomials in several variables: a polynomial in $\mathbb{D}[x_1,\ldots,x_n]$ can be viewed as a polynomial in one variable x_n (say) with coefficients in the domain $\mathbb{D}[x_1,\ldots,x_{n-1}]$. That is the main reason we went to the trouble of recalling the theory of polynomials in one variable *over a general domain*. Here is a useful fact about polynomials in several variables, which illustrates this philosophy well.

Lemma 3.9 *Let $f(x_1,\ldots,x_n)$ be a polynomial over an infinite domain $\mathbb{D}$ with the property that $f(x_1,\ldots,x_n) = 0$ for all $x_1, \ldots, x_n$ in $\mathbb{D}$. Then $f = 0$ in $\mathbb{D}[x_1,\ldots,x_n]$.*

Proof The proof proceeds by induction on the number n of variables. When $n = 1$ the result is immediate from Lemma 3.6. We proceed by induction, assuming the result holds for polynomials in $(n-1)$ variables. Write $f = f_0 + f_1 x_n + \cdots + f_s x_n^s$, where the coefficients f_k lie in the domain $\mathbb{D}[x_1,\ldots,x_{n-1}]$. If $f = 0$, there is nothing to prove, so we can suppose $f \neq 0$. Moreover, we can suppose $f_s \neq 0$, else we are back to the case of a polynomial in $(n-1)$ variables. By the induction hypothesis, there exist $a_1, \ldots, a_{n-1}$ in $\mathbb{D}$ for which $f_s(a_1,\ldots,a_{n-1}) \neq 0$. But then Lemma 3.6 tells us there are only finitely many values of x_n for which $f(a_1,\ldots,a_{n-1},x_n) = 0$. Choosing $a_n \in \mathbb{D}$ distinct from these values gives $f(a_1,\ldots,a_{n-1},a_n) \neq 0$, a contradiction. □

The *degree* of the monomial $x_1^{i_1} \ldots x_n^{i_n}$ is the natural number $i_1 + \cdots + i_n$, and that of the polynomial $f(x_1,\ldots,x_n)$ is the maximal degree of those monomials $x_1^{i_1} \dots x_n^{i_n}$ with $a_{i_1 \dots i_n} \neq 0$. As in the one variable case, it is easily checked that the degree is *additive*, in the sense that for polynomials $f_1, \dots, f_p$ the degree of the product $f = f_1 \dots f_p$ is given by

$$\deg f = \deg f_1 + \cdots + \deg f_p.$$

Clearly, the degree of a polynomial provides a degree function on any polynomial domain. What is not so obvious is that *provided the domain is a field* irreducible polynomials are prime. We are not going to prove this result, the interested reader will find excellent accounts given in numerous texts in the area of algebraic geometry. (For instance *Ideals, Varieties and Algorithms* by D. Cox, J. Little and D. O'Shea, Springer Verlag, Undergraduate Texts in Mathematics.) On that assumption Lemma 3.3 yields one of the most basic results used in this text.

Theorem 3.10 *Any non-constant polynomial $f(x_1, \dots, x_n)$ over a field $\mathbb{K}$ can be written uniquely (up to order and non-zero scalars) in the form $f = cf_1^{r_1} \dots f_s^{r_s}$ where c is a scalar, where $f_1, \dots, f_s$ are irreducible, and for $i \neq j$ no f_i is a factor of f_j.* (Unique Factorization Theorem.)

The polynomials $f_1, \dots, f_s$ are the *components* of f, and the numbers $r_1, \dots, r_s$ are their *multiplicities*. Note that if $f, f_1, \dots, f_s$ have degrees $d, d_1, \dots, d_s$ then (by additivity of the degree) $d = r_1 d_1 + \cdots + r_s d_s$. A component of multiplicity ≥ 2 is said to be *repeated*. A general remark is that the zero set of a curve f, with unique factorization $f = cf_1^{r_1} \dots f_s^{r_s}$, is the union of the zero sets of its components $f_1, \dots, f_s$. Thus in some sense, one is reduced to the irreducible case. In general it is rather difficult to determine whether a given polynomial in two variables is irreducible. Polynomials of degree 1 are certainly irreducible; for those of higher degree one may be able to decide the question by going back to first principles.

Example 3.3 Consider the polynomial $f = y^2 - x$. Let us verify from first principles that f is irreducible. Suppose f reduces to two polynomials of degree 1, so we can write $y^2 - x = (a_1x + b_1y + c_1)(a_2x + b_2y + c_2)$ with one of a_1, b_1 non-zero, and one of a_2, b_2 non-zero. Equating coefficients of x^2, xy, y^2, x we get $0 = a_1a_2$, $0 = a_1b_2 + a_2b_1$, $1 = b_1b_2$, $-1 = a_1c_2 + a_2c_1$. The first relation tells us that $a_1 = 0$ or $a_2 = 0$; we claim that *both* are zero. If $a_1 = 0$ the second relation yields $0 = a_2b_1$, but the third relation shows that $b_1 \neq 0$, so $a_2 = 0$ as well; likewise if $a_2 = 0$ then $a_1 = 0$ as well. The fourth relation now provides a contradiction, and we conclude that f is irreducible.

A few words are in order on the factorization of real and complex polynomials. In many physical problems one starts with a curve f in $\mathbb{R}^2$, and finds that the geometry is explained by thinking of f as a curve in $\mathbb{C}^2$ with real coefficients. Thus real and complex curves play a significant role in the applications of the subject. Recall from school mathematics that the factors of a polynomial in *one* variable with real coefficients are either real, or occur in complex conjugate pairs. This statement extends to general polynomials as follows. A complex polynomial f is said to be *real* when there exists a polynomial g *with real coefficients* and a complex scalar λ for which $f = \lambda g$. In particular, any polynomial f with real coefficients is real. (Beware: according to this definition the complex polynomial ix is real.) An equivalent condition proceeds via the idea of 'complex conjugation'. The *complex conjugate* of f is defined to be the polynomial $\bar{f}$ obtained from f by replacing all its coefficients by their complex conjugates. (For instance the complex conjugate of $x + iy$ is $x - iy$.) It is easily verified (Exercise 3.3.4) that f is real if and only if $f = \lambda\bar{f}$ for some complex number $\lambda \neq 0$. Note that f is irreducible if and only if $\bar{f}$ is irreducible; it follows immediately that if f has components $f_1, \dots, f_s$ then $\bar{f}$ has components $\bar{f}_1, \dots, \bar{f}_s$ with the same multiplicities.

Lemma 3.11 *The components of a real polynomial f are either real, or occur in complex conjugate pairs.*

Proof Suppose that f is real, so $f = \lambda\bar{f}$ for some complex number $\lambda \neq 0$. Then (by the Unique Factorization Theorem) f, $\bar{f}$ have the same components, and each f_j is a complex scalar multiple of some $\bar{f}_k$. When $j = k$ that means that f_j is real, and when $j \neq k$ it means that $\bar{f}_k$ is the complex conjugate of f_j, up to a complex scalar multiple. □

Exercises

3.3.1 Show from first principles that the conic x^2+y^2-1 is irreducible.

3.3.2 Show that $\deg(f+g) \leq \max(\deg f, \deg g)$.

3.3.3 Show that the degrees of polynomials f, g are related to the degree of the polynomial fg by $\deg(fg) = \deg f + \deg g$. Now use induction to establish that the degree is additive.

3.3.4 Show that a complex polynomial f is real if and only if there exists a complex number λ such that $f = \lambda\bar{f}$.

3.3.5 Show that any non-constant polynomial $f(x)$ over the field $\mathbb{R}$ has a unique factorization into linear factors $(x - \alpha)$ and quadratic factors $x^2 + \beta x + \gamma$ where α, β, γ are real. (Use Lemma 3.11.) Deduce that any non-constant polynomial $f(x)$ over the field $\mathbb{R}$ of *odd* degree has at least one real zero.

3.3.6 Find the zeros of the complex polynomial $f(x) = x^4 + 81$. Hence write $f(x)$ as a product of complex linear factors, and as a product of real quadratic factors.

3.3.7 Show that $x = i - 1$ is a zero of the complex polynomial given by $f(x) = x^4 + 2x^3 + 3x^2 + 2x + 2$. Hence factorize $f(x)$, and find all the zeros.

3.4 Homogeneous Polynomials

A polynomial F of degree d in the variables $x_1, \ldots, x_n$ over the domain $\mathbb{D}$, is *homogeneous* (or a *form*) when every monomial has the *same* degree d. Forms will be denoted by capital letters F, G, H, $\ldots$. Forms of degree 1, 2, 3, ... are said to be *linear*, *quadratic*, *cubic*, $\ldots$. Collecting together terms of given degree we see that any non-zero polynomial f of degree d can be written uniquely as $f = F_0 + F_1 + \cdots + F_d$ where each F_k is a form of degree k. One refers to F_0, F_1, $\ldots$, F_d as the *constant*, *linear*, *quadratic*, ... parts of f. We refer to the part F_c of lowest degree as the *lowest order terms* (abbreviated to LOT), and to the part F_b of highest degree as the *highest order terms* (abbreviated to HOT) in f. We will be interested mainly in *binary* forms (the case of two variables) and *ternary* forms (the case of three variables). Here is an alternative way of viewing forms. In practice we will only use it for the real and complex fields, but it is phrased for general infinite domains $\mathbb{D}$ simply to clarify the structure of the proof.

Lemma 3.12 *Let $F(x_1, \ldots, x_n)$ be a polynomial of degree d over an infinite domain $\mathbb{D}$. Then F is a form if and only if for all $t \in \mathbb{D}$,*

$$F(tx_1, \ldots, tx_n) = t^d F(x_1, \ldots, x_n). \tag{3.1}$$

Proof Equation 3.1 certainly holds when F is a form of degree d: multiplying all the variables by t means that we multiply each monomial (and hence F) by t^d. Conversely, assume Equation 3.1 holds. Write $F = F_0 + F_1 + \cdots + F_d$ where each F_k is a form of degree k. Since Equation 3.1 holds for each form F_k it can be rewritten as

$$F_0 + tF_1 + \cdots + t^d(F_d - F) = 0.$$

This is a polynomial equation in the single variable t with coefficients in the domain $\mathbb{D}[x_1,\ldots,x_n]$ having infinitely many roots, one for each $t \in \mathbb{D}$. By Lemma 3.6 all the coefficients vanish, so $F_0 = 0, \ldots, F_{d-1} = 0$, $F = F_d$. Thus F is a form of degree d. □

Since forms are special types of polynomials, they enjoy the property of unique factorization; what is not immediately clear is that factors of forms are themselves necessarily forms.

Lemma 3.13 *Any factor of a form F is itself a form.*

Proof Suppose $F = fg$, and that one (say f) of f, g is not a form. Write $f = F_a + F_{a+1} + \cdots + F_b$ with the F_i forms of degree i, and $F_a \neq 0$, $F_b \neq 0$. Also write $g = G_c + G_{c+1} + \cdots + G_d$ with the G_j forms of degree j, and $G_c \neq 0$, $G_d \neq 0$. Then $F = F_aG_c + \cdots + F_bG_d$ with $F_aG_c \neq 0$, $F_bG_d \neq 0$. By supposition $a < b$ so $\deg(F_aG_c) = a + c < b + d = \deg(F_bG_d)$, and that contradicts the assumption that F is a form. □

Thus, by Unique Factorisation of Polynomials, any form F admits a unique factorization $F = cF_1^{r_1}\ldots F_s^{r_s}$ with $c \neq 0$ a scalar, and the F_i irreducible forms such that F_j is not a factor of F_k for $j \neq k$. We will have a particular interest in binary forms, which have particularly simple factorizations.

Lemma 3.14 *Any non-zero complex form $F(x,y)$ can be written uniquely (up to order) in the form $F(x,y) = (\alpha_1 x + \beta_1 y)^{r_1}\ldots(\alpha_s x + \beta_s y)^{r_s}$ where the ratios $(\alpha_1 : \beta_1), \ldots, (\alpha_s : \beta_s)$ are uniquely defined.*

Proof Let $F(x,y) = \sum_{i=0}^{d} a_i x^i y^{d-i}$, and let c be the largest index for which $a_c \neq 0$. Then, using the unique factorization of polynomials in a single variable x/y for the second line, we derive the required form

$$\begin{aligned} F(x,y) &= y^d \sum_{i=0}^{c} a_i \left(\frac{x}{y}\right)^i \\ &= a_c y^d \left(\frac{x}{y} - \gamma_1\right)^{r_1} \cdots \left(\frac{x}{y} - \gamma_s\right)^{r_s} \\ &= a_c y^{d-c}(x - \gamma_1 y)^{r_1}\ldots(x - \gamma_s y)^{r_s}. \end{aligned}$$

□

Finally, here is a useful little proposition, providing us with a sufficient condition for the irreducibilty of many of the curves arising in this book.

Lemma 3.15 *Let F, G be non-zero forms of degrees d, $d+1$ respectively, having no common factors. Then $f = F + G$ is irreducible.*

Proof Suppose $F + G = ab$ with a, b non-constant polynomials. Write $a = A_p + \cdots + A_{p'}$, $b = B_q + \cdots + B_{q'}$ where the A_i, B_j are forms of degrees i, j where $p \leq p', q \leq q'$, and where A_p, $A_{p'}$, B_q, $B_{q'}$ are all $\neq 0$. Suppose $p < p'$ and $q < q'$. Equating terms of equal degree in $F + G = ab$ we get $F = A_pB_q$, $G = A_pB_{q+1} + A_{p+1}B_q, \ldots, 0 = A_{p'}B_{q'}$. The last relation is a contradiction, so we must have $p = p'$ or $q = q'$. But in either case it is clear that F, G have a non-constant common factor, contradicting the hypothesis. □

Example 3.4 The real polynomial $f = (x^2 - y^2)^2 - y(3x^2 - y^2)$ can be written $f = F_3 + F_4$, where $F_3 = -y(3x^2 - y^2) = -y(\sqrt{3}x - y)(\sqrt{3}x + y)$, $F_4 = (x^2 - y^2)^2 = (x-y)^2(x+y)^2$. Clearly F_3, F_4 have no common factors, so f is irreducible.

Exercises

3.4.1 Show that the criterion for homogeneity given in Lemma 3.12 may fail over a finite domain $\mathbb{D}$ by considering polynomials over the field $\mathbb{Z}_2$.

3.4.2 In each of the following cases, find the linear components of the given complex binary form $F(x, y)$.

(i) $F = x^2 + xy + y^2$ (iv) $F = x^3 + x^2y + xy^2 + y^3$
(ii) $F = x^3 - y^3$ (v) $F = x^4 + y^4$
(iii) $F = x^3 + y^3$ (vi) $F = x^4 + x^2y^2 + y^4$.

3.4.3 In each of the following cases show that the given polynomial $f(x, y)$ is irreducible.

(i) $f = y^2 - xy^2 - x^2 - x^3$
(ii) $f = y^3 - y^2 + x^3 - x^2 + 3xy^2 + 3x^2y + 2xy$
(iii) $f = x^2y - y^3(2x + y) - \lambda x^4 \quad (\lambda \neq 0)$
(iv) $f = (x^2 + y^2)^2 - y(3x^2 - y^2)$
(v) $f = x^6 - x^2y^3 - y^5$.

3.5 Formal Differentiation

We need to say something (it will be the bare minimum) about differentiating polynomial functions, since many of our results (and computations) will involve derivatives. The *formal derivative* of a polynomial $f(x) = a_0 + a_1x + a_2x^2 + \cdots$ is defined to be $f'(x) = a_1 + 2a_2x + 3a_3x^2 + \cdots$. The point of this definition is that it makes sense over any field, and gives the 'correct' answers (meaning those expected from elementary calculus) over the real (or for that matter, complex) field. In exactly the same way, we define the formal partial derivatives of a polynomial f in several variables, and the formal higher derivatives. We write f_x, f_y, ... for the first order partial derivatives with respect to the variables x, y, ... , and extend this notation in an obvious way to higher order derivatives. It is an easy matter to check that the resulting process of differentiation has the algebraic properties expected of differentiation. (Exercise 3.5.1.) Here is a typical application of derivatives to the question of when a polynomial $f(x)$ has a repeated factor.

Lemma 3.16 *A polynomial f in a single variable x over a field $\mathbb{K}$ has a repeated irreducible factor g if and only if g is a common factor of both f, f'.*

Proof Suppose that f has a repeated irreducible factor g, so $g^2|f$ and we can write $f = g^2h$ for some polynomial h. Differentiation yields $f' = 2gg' + g^2h'$, so g is a factor of f' as well. Conversely, suppose that g is an irreducible common factor of f, f'. Since $g|f$ we can write $f = gk$ for some polynomial k, and differentiation gives $f' = g'k + gk'$. As $g|f'$ we deduce that $g|g'k$. g is irreducible by assumption, hence prime. (Remember, it is a working assumption that in polynomial domains irreducibles are prime.) It follows that $g|g'$ or $g|k$; the former case is not possible, since $\deg g > \deg g'$, so $g|k$. Together with $f = gk$ that implies that $g^2|f$, so g is a repeated factor of f. □

The key application of differentiation is the idea of a Taylor expansion. This has two ingredients, namely the Taylor expansion for the special case of polynomials in one variable t, and the Chain Rule.

Lemma 3.17 *Let $\phi(t)$ be a polynomial of degree d over the field $\mathbb{K}$, and let a be a scalar. Then we have the following Taylor expansion of $\phi(t)$ centred at $t = a$,*

$$\phi(t) = \phi(a) + (t-a)\phi'(a) + \frac{1}{2}(t-a)^2\phi'(a) + \cdots + \frac{1}{d!}(t-a)^d\phi^{(d)}(a).$$

Proof Write $\phi(t+a) = \sum_{i=0}^{d} a_i t^i$. Differentiating d times with respect to t, and then setting $t = 0$ in the resulting expressions, we find that

$$a_0 = \phi(a), \quad a_1 = \phi'(a), \quad a_2 = \frac{1}{2}\phi'(a), \quad \ldots, \quad a_d = \frac{1}{d!}\phi^{(d)}(a).$$

The result follows on substituting these expressions for the coefficients $a_0, \ldots, a_d$ in the formula for $\phi(t+a)$, and then replacing t by $t-a$. □

The most useful case of the Taylor expansion is the case when $a = 0$, yielding the *Maclaurin expansion*

$$\phi(t) = \phi(0) + t\phi'(0) + \frac{1}{2}t^2\phi''(0) + \cdots + \frac{1}{d!}t^d\phi^{(d)}(0).$$

Lemma 3.18 *Let $f(x_1,\ldots,x_n)$ be a polynomial in n variables, and let $g_1(x), \ldots, g_n(x)$ be n polynomials in one variable, then we have the Chain Rule*

$$f(g_1(x),\ldots,g_n(x))' = \sum f_{x_i}(g_1(x),\ldots,g_n(x))g_i'(x).$$

We are now in a position to extend the Maclaurin expansion to any number of variables; effectively, this is an explicit way of writing the polynomial as a sum of forms.

Lemma 3.19 *Let $f(x)$ be a polynomial in $x = (x_1,\ldots,x_n)$ over the field $\mathbb{K}$, then we have the Maclaurin expansion at $0 = (0,\ldots,0)$*

$$f(x) = f(0) + \sum x_i f_{x_i}(0) + \frac{1}{2!}\sum x_i x_j f_{x_i x_j}(0) + \ HOT\,.$$

Proof For fixed x write $\phi(t) = f(tx)$. Expand $\phi(t)$ by the Maclaurin expansion, as in Lemma 3.17. The Chain Rule gives $\phi(0) = f(0)$, $\phi'(0) = \sum x_i f_{x_i}(0), \ldots$. The result follows on substituting these expressions back in the Maclaurin expansion, and then setting $t = 1$. □

Exercise

3.5.1 Establish the following properties of differentiation for polynomials over an arbitrary field.

(i) $(f+g)' = f' + g'$
(ii) $(fg)' = f'g + fg'$
(iii) $f' = 0$ for f constant.

4
Affine Equivalence

In this chapter we will be concerned with the question of when two curves are to be regarded as the 'same'. The first thing to be clear about is that there is no absolute answer – it all depends on which aspects of curves you are studying. We will introduce the group of 'affine mappings', leading to a natural relation of 'affine equivalence' on the curves of given degree d. The remainder of the chapter will be concerned with concepts which are invariant under affine mappings, so can be used in principle to distinguish curves. The first of these is the concept of degree, which has a basic geometrical interpretation, and the second is the interesting concept of a 'centre', inspired by the familiar case of a circle.

In order to motivate the definitions of Section 4.1, it may be helpful to start with what is possibly a more familiar notion, at least in the context of studying the standard conics of elementary geometry. We keep to the familiar Euclidean plane $\mathbb{R}^2$. (This is one of the very few points in our development where we will refer to the Euclidean structure on $\mathbb{R}^2$, but only for motivational purposes.) When dealing with standard conics, one thinks of two curves as being 'rigidly' equivalent when the one can be superimposed exactly on the other. You think of the curves as being in different (rectangular) coordinate systems, say one curve in the 'standard' (x, y)-plane, and the other in the (X, Y)-plane. Then the instinct is that if you place the second plane on the first so that the axes coincide, then the two curves will coincide exactly. The idea is illustrated in Figure 4.1.

The relation between the two coordinate systems in this situation is that the one can be obtained from the other by a *rigid mapping* from the (x, y)-plane to the (X, Y)-plane, i.e. a mapping that preserves both distance and orientation. In books on linear algebra, it is shown that rigid maps are precisely those which can be written as composites of rotations (about the origin) and translations. Explicitly, they are mappings given

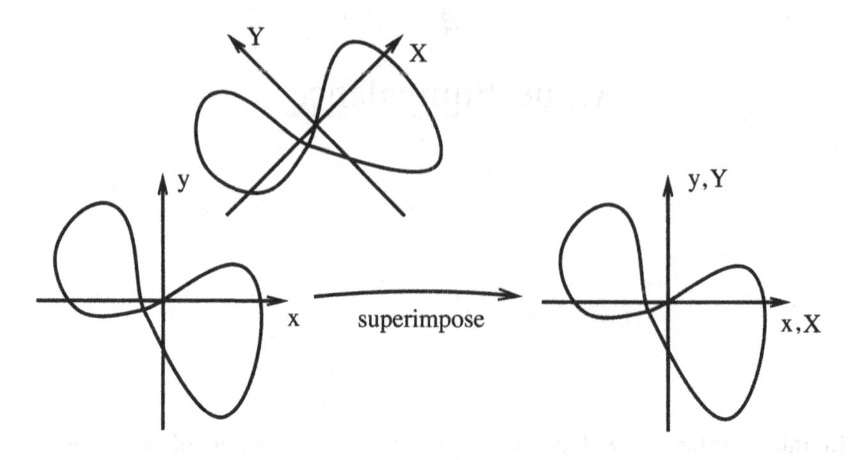

Fig. 4.1. Superimposing two curves

by formulas of the following form, where θ is the angle of rotation, and (u, v) is the translation vector.

$$\begin{cases} X = \cos\theta \cdot x - \sin\theta \cdot y + u \\ Y = \sin\theta \cdot x + \cos\theta \cdot y + v. \end{cases}$$

The resulting relation of 'rigid' equivalence on curves is too fine, in the sense that it produces far too many types of curves. For instance, two standard parabolas $y^2 = 4ax$, $y^2 = 4a'x$ with $a, a' > 0$ are rigidly equivalent in this sense if and only if $a = a'$. (Exercise 4.2.2.) Thus we would obtain infinitely many different types of parabola, one for each positive value of the scalar a. However, there is a natural way forward here. The rotational component of the 'rigid' map is a special type of invertible linear map; replacing this by a general invertible linear map we obtain the key concept of an 'affine' map, developed in the remainder of this chapter. This has the extra advantage that it will make sense over a general field $\mathbb{K}$.

4.1 Affine Maps

The first step is to introduce the concept of an 'affine map' of $\mathbb{K}^2$. An *affine* map (or equivalence, or change of coordinates) of $\mathbb{K}^n$ is a mapping $\phi : \mathbb{K}^2 \to \mathbb{K}^2$ of the form $\phi = \tau \circ \lambda$, where λ is an invertible linear map of $\mathbb{K}^2$, and τ is a 'translation' defined by a formula $\tau(x) = x + t$ with t a fixed vector in $\mathbb{K}^2$. Thus, if we think of λ given by a formula $\lambda(x) = Lx$, with L an invertible 2×2 matrix, we see that an affine map has the form $\phi(x) = Lx + t$. More concretely, an affine mapping has the form $\phi(x, y) = (X, Y)$ where X, Y are given by the formulas below for some scalars p, q, r, s, u, v with $ps - qr \neq 0$.

$$\begin{cases} X &= px + qy + u \\ Y &= rx + sy + v. \end{cases}$$

Example 4.1 Taking the linear map λ to be the identity, we see that any translation is an affine map. Likewise, taking the translation τ to be zero we see that any invertible linear map is affine. In particular, *scalings* $X = px$, $Y = sy$ (with $p \neq 0$, $s \neq 0$) are affine maps, as is the *coordinate switch* $X = y$, $Y = x$. Rigid mappings of the plane are affine, corresponding to the case when $p = \cos\theta$, $q = -\sin\theta$, $r = \sin\theta$, $s = \cos\theta$ for some angle θ; note that $ps - qr = 1 \neq 0$.

Recall that translations and invertible linear maps are examples of bijections of the plane. Since the composite of two bijections is another bijection, we deduce that affine maps are automatically bijective, and hence have inverses. One of the basic facts about affine maps of $\mathbb{K}^2$ is that they form a group $Aff(2)$ under the operation of composition. (It is *not* Abelian.) We need to verify the group axioms of Section 2.2.

- G1. $Aff(2)$ is closed under composition. Indeed if $\phi_1(z) = L_1z + t_1$, $\phi_2(z) = L_2z + t_2$ are two affine maps then (by straight computation) their composite is $\phi(z) = (\phi_1 \circ \phi_2)(z) = Lz + t$, where $L = L_1L_2$ and $t = L_1t_2 + t_1$, which is again an affine map. (Recall from linear algebra that the product of two invertible 2×2 matrices is again invertible.)
- G2. Composition of maps is known to be an associative operation on the set of all maps from a given set to itself.
- G3. The identity map of $\mathbb{K}^2$ is the affine map $\phi(z) = Lx + t$ with $L = I$, $t = 0$ where I is the identity 2×2 matrix.
- G4. $Aff(2)$ is closed under inversion. The inverse of the affine map $\phi(z) = Lz + t$ is $\phi^{-1}(z) = L'z + t'$ where $L' = L^{-1}$ and $t' = -L^{-1}t$, which is again affine.

4.2 Affine Equivalent Curves

Two affine curves f, g in $\mathbb{K}^2$ are *affinely equivalent* when there exists an affine map ϕ and a scalar $\lambda \neq 0$ with $g(x, y) = \lambda f(\phi(x, y))$. Since $Aff(2)$ is a group, affine equivalence defines an equivalence relation on

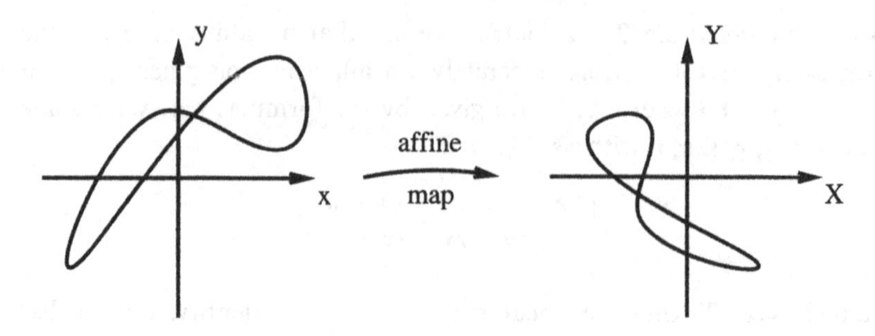

Fig. 4.2. Affine equivalent curves

the curves in $\mathbb{K}^2$. One pictures the situation as in Figure 4.2; two curves as the 'same' when the one can be superimposed upon the other by an affine map. More concretely, that means that there exist scalars p, q, r, s with $ps - qr \neq 0$ and a relation of the form

$$g(x, y) = \lambda f(px + qy + \alpha, rx + sy + \beta).$$

Example 4.2 Any two lines in $\mathbb{K}^2$ are affinely equivalent. Since the affine equivalences form a group, it is sufficient to show that any line is affinely equivalent to the line $x = 0$. Suppose the line is $ax + by + c = 0$, with one of a, b non-zero. We can suppose $a \neq 0$, otherwise we just switch the coordinates x, y to obtain an affine equivalent line with $a \neq 0$. Then $X = ax + by + c$, $Y = y$ is an affine equivalence, with inverse $x = (X - bY - c)/a$, $y = Y$, taking $ax + by + c$ to X.

Example 4.3 In $\mathbb{R}^2$ the standard parabola $y^2 = 4ax$ with $a > 0$ is affine equivalent to the parabola $Y^2 = X$ via the scaling $x = aX$, $y = 2aY$; the standard ellipse $x^2/a^2 + y^2/b^2 = 1$ with $0 < a < b$ is affinely equivalent to the circle $X^2 + Y^2 = 1$ via the scaling $x = aX$, $y = bY$; and the standard hyperbola $x^2/a^2 - y^2/b^2 = 1$ is affine equivalent to the rectangular hyperbola $X^2 - Y^2 = 1$.

Generally, a *parabola* in $\mathbb{R}^2$ is a conic affine equivalent to $Y^2 = X$: an *ellipse* is a conic affine equivalent to $X^2 + Y^2 = 1$; and a *hyperbola* is a conic affine equivalent to $X^2 - Y^2 = 1$. We will have rather more to say about the affine types of conics in Chapter 5.

Exercises

4.2.1 Show that the rigid mappings of $\mathbb{R}^2$ form a subgroup of the affine group $Aff(2)$.

4.2.2 Two affine curves f, g in $\mathbb{R}^2$ are *rigidly equivalent* when there exists a rigid map ϕ and a scalar $\lambda \neq 0$ with $g(x, y) = \lambda f(\phi(x, y))$. Show that the parabolas $y^2 = 4ax$, $y^2 = 4a'x$ in $\mathbb{R}^2$ with $a, a' > 0$ are rigidly equivalent if and only if $a = a'$.

4.3 Degree as an Affine Invariant

In this section, we will establish that the algebraic concept of the degree of a curve is an affine invariant, and that it admits an important geometric interpretation.

Lemma 4.1 *The degree of a curve is an affine invariant, i.e. two affine equivalent curves have the same degree. Thus affine maps take lines to lines, conics to conics, and so on.*

Proof Let f, g be affine equivalent curves, so there exists an affine map $\phi(x, y) = (px + qy + \alpha, rx + sy + \beta)$ and a scalar $\lambda \neq 0$ such that $g(x, y) = \lambda f(\phi(x, y)) = \lambda f(px + qy + \alpha, rx + sy + \beta)$. Assume $f(x, y) = \sum a_{ij} x^i y^j$, so

$$g(x, y) = \lambda \sum a_{ij}(px + qy + \alpha)^i (rx + sy + \beta)^j.$$

Each expression $px + qy + \alpha$, $rx + sy + \beta$ in parentheses is a polynomial of degree 1. (They cannot be constant, else the condition $ps - qr \neq 0$ fails.) By additivity of the degree, $(px + qy + \alpha)^i (rx + sy + \beta)^j$ has degree $i + j$. It follows immediately that $\deg f \geq \deg g$: since affine equivalence is a symmetric relation $\deg g \geq \deg f$, yielding the result. □

Lemma 4.2 *Let f, g be affine equivalent curves in $\mathbb{K}^2$. Suppose that f has components $f_1, \dots, f_s$ with multiplicities $r_1, \dots, r_s$ and degrees $d_1, \dots, d_s$. Then g has components $g_1, \dots, g_s$ where f_k, g_k are affinely equivalent, with the same multiplicities $r_1, \dots, r_s$ and the same degrees $d_1, \dots, d_s$. In particular, f is irreducible (reducible) if and only if g is irreducible (reducible).*

Proof Since f, g are affinely equivalent there exists an affine map ϕ and a scalar $\lambda \neq 0$ with $g(x, y) = \lambda f(\phi(x, y))$. The curve f has a unique factorization $f = cf_1^{r_1} \dots f_s^{r_s}$ with $c \neq 0$. Then

$$\begin{aligned} g(x,y) &= \lambda f(\phi(x,y)) \\ &= (\lambda c f_1^{r_1} \dots f_s^{r_s})(\phi(x,y)) \\ &= \lambda c f_1^{r_1}(\phi(x,y)) \dots f_s^{r_s}(\phi(x,y)) \\ &= \lambda c g_1^{r_1}(x,y) \dots g_s^{r_s}(x,y) \end{aligned}$$

where $g_k(x,y) = f_k(\phi(x,y))$ is affinely equivalent to f_k, and hence has the same degree d_k. In particular, the argument shows that f is irreducible (reducible) if and only if g is irreducible (reducible). It follows that $g_1, \dots,$ g_s are the components of g. □

Our object now is to show that the algebraic concept of degree has a purely geometric interpretation in terms of the zero set of the curve, providing a powerful way of thinking about curves. The proof depends on a useful technical lemma, which gives you some idea of the 'flexibility' of affine maps. A set of points in $\mathbb{K}^2$ is *collinear* when there exists a line passing through every point in the set. And a set of lines in $\mathbb{K}^2$ is *concurrent* when there exists a point through which every line in the set passes.

Lemma 4.3 *Let A, B, C and A', B', C' be triples of non-collinear points. There exists a unique affine map ϕ which takes A, B, C respectively to A', B', C'.* (The Three Point Lemma.)

Proof Since the affine maps form a group, it is enough to prove the result when $A = (1,0)$, $B = (0,1)$, $C = (0,0)$. We seek an affine map $\phi(X,Y) = (x,y)$, say $x = pX + qY + \alpha$, $y = rX + sY + \beta$ with $ps - qr \neq 0$ for which $\phi(1,0) = (p+\alpha, r+\beta) = A'$, $\phi(0,1) = (q+\alpha, s+\beta) = B'$, $\phi(0,0) = (\alpha,\beta) = C'$. It is clear these relations determine α, β, p, q, r, s (and hence the map ϕ) uniquely. If $ps - qr = 0$, the vectors (p,r), (q,s) would be linearly dependent, so (p,r), (q,s), $(0,0)$, and hence their translates A', B', C', would be collinear, a contradiction: thus $ps - qr \neq 0$. □

Here is an application of the Three Point Lemma. It is one of the guiding intuitions of the subject, and will play a very basic role in the remainder of this book.

Lemma 4.4 *Let f be an affine curve of degree $d \geq 1$ in $\mathbb{K}^2$, and let l be a line. Then either l meets f in $\leq d$ points, or l is a component of f. Thus* *for a curve f with no line components the degree d is an upper bound for the number of intersections of f with any line l.*

Proof Let l be the line through $a = (a_1, a_2)$ and $b = (b_1, b_2)$. Then l is parametrized as $x(t) = (1-t)a_1 + tb_1$, $y(t) = (1-t)a_2 + tb_2$ with $t \in \mathbb{K}$. The intersections of f and l are given by $\phi(t) = 0$ where $\phi(t) = f(x(t), y(t))$ is a polynomial in t of degree $\leq d$ (possibly $< d$). Assume ϕ is not identically zero, then the equation $\phi(t) = 0$ has $\leq d$ roots, and there are $\leq d$ points of intersection. If on the other hand ϕ is identically zero, then f vanishes at every point on l. By the Three Point Lemma we can assume $a = (0,0)$, $b = (1,0)$, then l is given by $y = 0$, and f vanishes on l if and only if $f(x,0) = 0$ for all $x \in \mathbb{K}$. Set $f(x,y) = yg(x,y) + h(x)$ with g, h polynomials, then $h(x) = 0$ for all $x \in \mathbb{K}$; that means $h = 0$, so y is a factor of $f(x,y)$, i.e. l is a component of f. □

In particular, a line is either a component of an algebraic curve or meets it in *finitely* many points. (This is one of the 'finiteness' properties of algebraic curves mentioned in Section 1.1. It allows us to give simple examples of curves which fail to be algebraic. For instance the graph of the sine function $y = \sin x$ is not algebraic, since the line $y = 0$ intersects it in infinitely many points, but is not a component.)

Exercises

4.3.1 Show that the polynomial $f = 3xy + x^3 + y^3$ is irreducible. Deduce that $g = x^3 + y^3 - 3x^2 - 3y^2 + 3xy + 1$ is also irreducible. (Hint: it suffices to show that f, g are affine equivalent.)

4.4 Centres as Affine Invariants

The idea of a 'centre' of a curve is motivated by the familiar case of a circle in $\mathbb{R}^2$. Any line through the centre meets the circle in two distinct points, and the centre is the mid-point of the resulting line segment. What is interesting is that curves other than the circle can have 'centres' in this sense, and that the idea is invariant under affine mappings. A formal development is based on the following observation, that in a certain sense affine maps preserve parameters on lines.

Lemma 4.5 *Let a, b be distinct points in $\mathbb{K}^2$, let $t \in \mathbb{K}$ and let $c = a + t(b-a)$. Then, for any affine map f, we have $f(c) = f(a) + t(f(b) - f(a))$.*

Thus an affine map f maps the parametrized line through a, b to the parametrized line through $f(a)$, $f(b)$, and 'preserves parameters'.

Proof For any vector $u \in \mathbb{K}^2$ write $f(u) = \lambda(u) + k$, with λ linear, and k a fixed vector in the plane. Then, using linearity of λ to get the third line,

$$\begin{aligned} f(c) &= f(a + t(b-a)) \\ &= \lambda(a + t(b-a)) + k \\ &= \lambda(a) + t(\lambda(b) - \lambda(a)) + k \\ &= \{\lambda(a) + k\} + t(\lambda(b) - \lambda(a)) \\ &= f(a) + t\{(\lambda(b) + k) - (\lambda(a) + k)\} \\ &= f(a) + t(f(b) - f(a)). \end{aligned}$$

By definition, the parametrized line through a, b comprises the set of points $a+t(b-a)$, and the parametrized line through $f(a)$, $f(b)$ comprises the set of points $f(a) + t(f(b) - f(a))$, so (by the above computation) comprises the image under f of the parametrized line through a, b. □

Example 4.4 The *mid-point* of the parametrized line through two points a, b in $\mathbb{K}^2$ is defined to be the point with parameter $t = 1/2$, i.e. the point $c = \frac{1}{2}(a + b)$. Then Lemma 4.5 tells us that affine maps preserve mid-points on parametrized lines: under an affine map f the mid-point c of the line-segment joining a, b is mapped to the mid-point $f(c)$ of the line segment joining $f(a)$, $f(b)$.

Example 4.5 Affine maps preserve the property of two subsets of the plane intersecting, in the sense that if A, B are subsets of the plane, and f is an affine map, then A, B intersect if and only if $f(A)$, $f(B)$ intersect; indeed this is a property of general bijective maps. In particular, the concept of 'parallelism' for lines is an affine invariant.

That brings us to the key definition of this section. Let $p = (\alpha, \beta)$ be a fixed point in $\mathbb{K}^2$. By *central reflexion* in the point p we mean the affine mapping defined by the formula $(x, y) \mapsto (x', y')$ where $x' = 2\alpha - x$, $y' = 2\beta - y$. (Note that when p is the origin this simply means that $x' = -x$, $y' = -y$.) The geometric content is that the mid-point of the line joining a point (x, y) to its central reflexion (x', y') in p is the point p. Now let f be a curve in $\mathbb{K}^2$. We say that p is a *centre* for f when the curves $f(x, y)$, $f(x', y')$ are the same, i.e. their defining polynomials differ only by a non-zero scalar multiple. We express this by saying that f is invariant under central reflexion in the point p. A curve f is *central* when it has at least one centre.

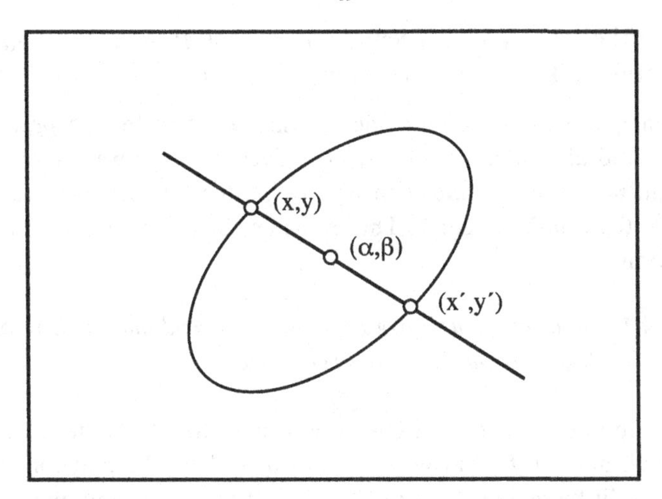

Fig. 4.3. The concept of a centre

Example 4.6 Any line $ax + by + c$ is a central curve. Indeed any point $p = (\alpha, \beta)$ on the line is a centre, for $ax' + by' + c = a(2\alpha - x) + b(2\beta - y) + c = 2(a\alpha + b\beta + c) - (ax + by + c) = -(ax + by + c)$.

Example 4.7 Consider a conic which reduces to a pair of lines l_1, l_2. A point p of intersection is called a *vertex* of the conic. (Thus when the lines are distinct and parallel there is no vertex, when the lines are distinct and intersect there is a unique vertex, and when the lines coincide there is a line of vertices.) A vertex p is automatically a centre of the conic: by Example 4.6 each line l_1, l_2 is invariant under central reflexion in p, and hence the product $l_1 l_2$ is invariant as well.

Example 4.8 The conic $y^2 = 1$ defines parallel lines $y = \pm 1$. Any point $p = (\alpha, 0)$ is a centre for the conic. Indeed for any point (x, y), the central reflexion in the point p is the point (x', y') where $x' = 2\alpha - x$, $y' = -y$, and clearly (x, y) lies on the conic if and only if (x', y') does. Thus the line $y = 0$, parallel to $y = \pm 1$ and 'midway' between them, provides a line of centres.

Lemma 4.6 *Centres of curves are affine invariant, in the following sense. Let p be a centre for a curve f, and let ϕ be an affine mapping, then $p' = \phi^{-1}(p)$ is a centre for $f' = f \circ \phi$. In particular, if two curves are affine equivalent, and one is central, then the other is central.*

Proof Write $z = (x, y)$. Then $f'(z) = f(\phi(z)) = f(2p - \phi(z)) = f(\phi(2\phi^{-1}(p) - z)) = f'(2\phi^{-1}(p) - z) = f'(2p' - z)$. □

In principle, Lemma 4.6 provides a starting point for the process of trying to find all centres $p = (\alpha, \beta)$ of a given curve f; we can translate that point to the origin, and then try to discover whether the origin is a centre for the translated curve. That provides the motivation for the next proposition.

Lemma 4.7 *The origin is a centre of a curve f if and only if there are no terms of odd degree, or no terms of even degree.*

Proof Suppose that the origin is a centre for f. Write $f(x, y) = \sum_{i,j} a_{ij}x^iy^j$, and let $f'(x, y) = \sum_{i,j}(-1)^{i+j}a_{ij}x^iy^j$ be the polynomial obtained from it by replacing x, y by $-x$, $-y$. The supposition implies that $f' = \lambda f$ for some scalar $\lambda \neq 0$, so that $a_{ij} = \lambda(-1)^{i+j}a_{ij}$ for all choices of i, j. It follows that $1 = \lambda(-1)^{i+j}$ for all choices of i, j with $a_{ij} \neq 0$. Since at least one $a_{ij} \neq 0$, that means that either all the $i + j$ with $a_{ij} \neq 0$ are even (and $\lambda = 1$), or they are all odd (and $\lambda = -1$). Conversely, if every term in f is of odd degree then $f' = -f$, and the origin is a centre; likewise, if every term in f is of even degree then $f' = f$, and we deduce again that the origin is a centre. □

Thus if f has even degree, the origin is a centre if and only if there are no terms of odd degree. The simplest case is provided by conics. In that case the origin is a centre if and only if the terms of degree 1 in x, y are absent. (For instance, the origin is a centre of the circle $x^2 + y^2 - 1$, of the hyperbola $x^2 - y^2 - 1$, and of the line-pair $x^2 - y^2$.) This observation provides the basis for a practical method of finding all the centres (α, β) of a given conic

$$ax^2 + 2hxy + by^2 + 2gx + 2fy + c. \qquad (\star)$$

Translate the centre to the origin by substituting $x = X + \alpha$, $y = Y + \beta$ and then see if $(0, 0)$ is a centre for the translated conic, by setting the coefficients of X, Y equal to zero. That yields two linear equations in α, β.

$$\begin{aligned}\text{coefficient of } X &= 2(a\alpha + h\beta + g) = 0\\ \text{coefficient of } Y &= 2(h\alpha + b\beta + f) = 0.\end{aligned}$$

Note that these equations are obtained by differentiating $(\star)$ with respect to x, y, writing $x = \alpha$, $y = \beta$, and then setting the results equal to zero. Geometrically, each equation defines a line in the (α, β)-plane. We now have the situation of Example 1.2. In general the lines intersect in a unique centre, but exceptionally they could be parallel (in which case there is no centre) or coincide (in which case there is a line of centres). These are the only possibilities. By linear algebra, there is a unique centre if and only if $\delta = ab - h^2 \neq 0$. (δ is the determinant of the coefficients.) This gives you a simple numerical test for determining whether or not a general conic ($\star$) has a unique centre. Note, however, that when $\delta = 0$ it does not distinguish the case when there is no centre from the case when there is a line of centres.

Example 4.9 Consider the conic $x^2-2xy+5y^2+2x-10y+1$. Here $a = 1$, $b = 5$, $h = -1$ so $\delta = 1.5 - (-1)^2 = 4 \neq 0$, and there is a unique centre. We can find it explicitly as follows. Substituting $x = X + \alpha$, $y = Y + \beta$, we need to solve the following equations. These have the solution $\alpha = 0$, $\beta = 1$ yielding the centre $(0, 1)$.

$$\begin{aligned} \text{coefficient of } X &= 2(\alpha - \beta + 1) &= 0 \\ \text{coefficient of } Y &= 2(-\alpha + 5\beta - 5) &= 0. \end{aligned}$$

Example 4.10 Consider the parabola $y^2 - x$ in $\mathbb{R}^2$. Substitute $x = X + \alpha$, $y = Y + \beta$ to see that the coefficients of X, Y are respectively -1, 2β. The condition for (α, β) to be a centre is that $-1 = 0$, $2\beta = 0$. We cannot satisfy these relations, so the parabola has no centre. It follows from Lemma 4.6 that the parabola cannot be affinely equivalent to a real ellipse, since the latter is central. This is an example where we can use an affine invariant to distinguish two (irreducible) affine curves of the same degree.

Here is a pretty application of the above ideas, making the point that they can readily establish facts which are by no means obvious. We have already seen that the conic $x^2 + y^2 - 1$ in $\mathbb{R}^2$ has a unique centre (not on the conic), and hence any conic affinely equivalent to it, i.e. any real ellipse, has a unique centre (not on the conic). By a *chord* of a real ellipse we mean a line which meets the conic in two distinct points, and by a *diameter* we mean a chord through the centre. Clearly, an affine map between two real ellipses will map any chord (respectively diameter) of the one to a chord (respectively diameter) of the other.

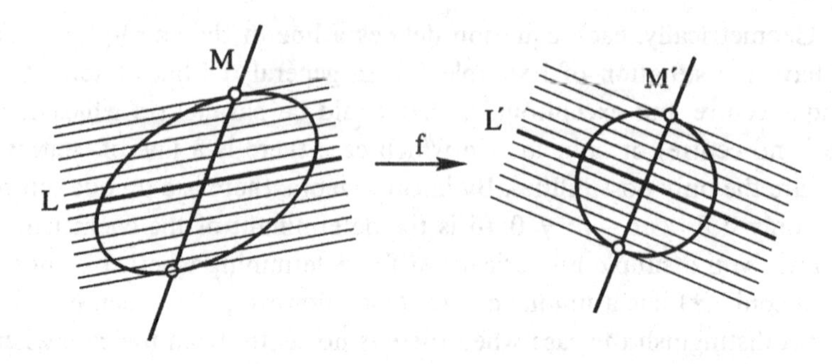

Fig. 4.4. Diameters of an ellipse

Lemma 4.8 *If you take all the chords parallel to a given diameter L of an ellipse E then the mid-points lie on another diameter M. (L, M are said to be 'conjugate diameters' of the ellipse.)*

Proof This follows almost immediately from the facts that parallelism and mid-points are preserved by affine maps. There is an affine map f taking E to the unit circle, and mapping the diameter L to a diameter L' of the circle. Clearly, the mid-points of all the chords parallel to L' lie on a diameter M' of the circle. And under the inverse f^{-1}, the diameter M' is mapped to a diameter M of E with the desired properties. (Figure 4.4.) □

Lemma 4.8 makes a fundamental point, namely that we can deduce affine properties of affine algebraic curves from those of curves affine equivalent to them. That raises the question of classifying curves of given degree d up to affine equivalence. Example 4.2 solves that problem in the simplest case $d = 1$. In Chapter 5, we will deal with the next simplest case $d = 2$ of conics in some detail, as they represent a key element in understanding algebraic curves.

Exercises

4.4.1 In each of the following cases decide whether the given conic in $\mathbb{R}^2$ has a centre, and if so write down the equation of the conic relative to the centre as origin.

(i) $25x^2 + 9y^2 - 72y - 81$

(ii) $5x^2 + 6xy + 5y^2 - 4x + 4y - 4$

(iii) $x^2 - 4xy + 4y^2 + 10x - 8y + 13$.

4.4.2 In each of the following cases decide whether the given conic in $\mathbb{R}^2$ has a centre, and if so, write down the equation of the conic relative to the centre as origin.

(i) $x^2 + xy - 2y^2 + x - y$
(ii) $x^2 - 4xy + 3y^2 + 2x$
(iii) $x^2 - 2xy + y^2 + 2x - 4y + 3$
(iv) $9x^2 - 24xy + 16y^2 + 3x - 4y - 2.$

5
Affine Conics

This chapter can be thought of as an extended example. The object is to introduce the reader to the question of listing (and of distinguishing) affine curves of given degree d under the relation of affine equivalence. For general values of d, this is a somewhat hopeless (and ultimately uninteresting) task. The answer in the simplest possible case $d = 1$ of lines (Example 4.2) is misleading, in the sense that it does not depend on the underlying field $\mathbb{K}$. The next case $d = 2$ of conics is more interesting, since part of the listing process applies over virtually any field $\mathbb{K}$, and beyond that it depends crucially on $\mathbb{K}$. It is particularly satisfying that we obtain complete answers over the real and the complex fields; in those cases we will manage to reduce an arbitrary conic to one of a finite list of 'normal forms' by affine changes of coordinates. By contrast, it would take a significant amount of number theory to achieve an answer over the rational field.

The conic case also forces us to think about the question of distinguishing two types of conic on our list. One basic distinction between lines and conics is that *conics may factorize* so the multiplicities of the components will provide affine invariants. For instance $x^2 - y^2$ factorizes as $(x - y)(x + y)$; it is an example of a 'line-pair' since its zero set is the union of the zero sets of the lines $x - y$, $x + y$. A more degenerate case still is provided by x^2, a 'repeated line'. However, there are certainly conics which do not reduce, for instance in Example 3.3 we saw that the parabola $f = y^2 - x$ is irreducible. In Section 5.2 we will look at more subtle invariants of conics, the so-called 'delta invariants' which enable us to make finer distinctions than those provided by the multiplicities of the components.

5.1 Affine Classification

The first step in the process of listing affine conics is to reduce to a 'prenormal' form, *independent of the field* $\mathbb{K}$. The resulting lists will depend on $\mathbb{K}$. According to the definition, a general conic in $\mathbb{K}^2$ is a polynomial of degree 2. In this chapter we will tacitly assume that the ground field $\mathbb{K}$ has characteristic $\neq 2$, so any conic can be written in the following form, where the coefficients a, b, c, f, g, h lie in $\mathbb{K}$, and at least one of a, b, h is non-zero,

$$ax^2 + 2hxy + by^2 + 2gx + 2fy + c. \qquad (\star)$$

Lemma 5.1 *Any conic in* $\mathbb{K}^2$ *is affinely equivalent to a prenormal form* $y^2 = q(x)$ *where* $q(x) = Ax^2 + 2Bx + C$ *and* A, B, C *are in* $\mathbb{K}$.

Proof The idea is to view $(\star)$ as a quadratic in y, whose coefficients involve x, and then 'complete the square'. For this to work we assume (for a moment) that $b \neq 0$, then, dividing through by b, we can suppose $b = 1$. That yields an equation $(y + hx + f)^2 = q(x)$, with $q(x)$ having the desired form. Now replace $y + hx + f$ by y, and leave x unchanged (an affine equivalence) to achieve the form $y^2 = q(x)$. If $b = 0$ and $a \neq 0$ just switch x, y (another affine equivalence) and proceed as before. That leaves us with the case $a = b = 0$, so $h \neq 0$. The trick is to replace x, y by $x + y$, $x - y$ (yet another affine equivalence) to get back to the case $a \neq 0$, $b \neq 0$. □

Next, let us see what this gives for the real and complex fields. (For other fields the list may be very difficult to obtain.) Here is the list of normal forms for real conics.

Lemma 5.2 *In* $\mathbb{R}^2$ *any conic is affinely equivalent to one of the nine normal forms in Table 5.1.*

Proof By Lemma 5.1, we can suppose the conic has the form $y^2 = Ax^2 + 2Bx + C$, where A, B, C are real numbers. Assume $A \neq 0$; scaling x by $\lambda \neq 0$ the coefficient of x^2 becomes $A\lambda^2$, and we can force this to be ± 1 by choosing λ so that $A\lambda^2 = \pm 1$. (When A is positive we take $+$, and when A is negative we take $-$.) Thus we can suppose $A = \pm 1$, so $y^2 = \pm(x - B)^2 + (B^2 - C)$, with possibly different B and C. (When A is negative change B into $-B$ as well.) Replacing $x - B$ by x, the equation assumes the form $y^2 = \pm x^2 + D$. When $D \neq 0$ we can scale both x, y by

Table 5.1. *The nine types of real affine conic*

normal form	name
$y^2 = x$	parabola
$y^2 = -x^2 + 1$	real ellipse
$y^2 = -x^2 - 1$	imaginary ellipse
$y^2 = x^2 + 1$	hyperbola
$y^2 = x^2$	real line-pair
$y^2 = -x^2$	imaginary line-pair
$y^2 = 1$	real parallel lines
$y^2 = -1$	imaginary parallel lines
$y^2 = 0$	repeated line

$\mu \neq 0$, and then divide through by μ^2, to change D to D/μ^2; now choose μ so that $D/\mu^2 = \pm 1$ to get $D = \pm 1$. (When D is positive we take +, and when D is negative we take −.) Firstly, that yields the two normal forms $y^2 = -x^2 \pm 1$ (the *real ellipse* and the *imaginary ellipse*). It also yields $y^2 = x^2 \pm 1$; note however that these forms are affinely equivalent, we just need to switch x and y, so we only get one normal form $y^2 = x^2 + 1$ representing the *hyperbola*. When $D = 0$ we get the *real line-pair* $y^2 = x^2$ and the *imaginary line-pair* $y^2 = -x^2$. Now consider the case $A = 0$, so $y^2 = 2Bx + C$. Suppose $B \neq 0$. Then we can replace $2Bx + C$ by x to get the *parabola* $y^2 = x$. When $B = 0$ our equation is $y^2 = C$. When $C \neq 0$ we can scale y by v, and then divide through by v^2 to replace C by C/v^2. Choose v so that $C/v^2 = \pm 1$. (When C is positive take the + sign, and when C is negative take the − sign.) Our equation is now $y^2 = \pm 1$, yielding *real parallel lines* $y^2 = 1$ and *imaginary parallel lines* $y^2 = -1$. Finally, when $C = 0$ we obtain the *repeated line* $y^2 = 0$. □

Example 5.1 Consider the conic $x^2 + 2xy + 5y^2 - 2x - 10y + 1$. Collecting terms in x and 'completing the square' we get $(x + y - 1)^2 + 4(y^2 - 2y)$, and 'completing the square' again gives $(x + y - 1)^2 + 4(y - 1)^2 - 4$. Under the affine equivalence $X = \frac{1}{2}(x + y - 1)$, $Y = y - 1$ this becomes $4(X^2 + Y^2 - 1)$, which is a *real ellipse*.

Trivial modifications to the proof of Lemma 5.3 (essentially omitting the various $\pm$ possibilities) yields the affine classification of complex conics.

Table 5.2. *The five types of complex affine conics*

normal form	name
$y^2 = x$	parabola
$y^2 = x^2 + 1$	general conic
$y^2 = x^2$	line-pair
$y^2 = 1$	parallel lines
$y^2 = 0$	repeated line

Lemma 5.3 *In $\mathbb{C}^2$, any conic is affinely equivalent to one of the five normal forms in Table 5.2.*

Exercises

5.1.1 In each of the following cases, show (by explicit factorization) that the given conic in $\mathbb{R}^2$ is a line-pair, and determine the vertex, i.e. the intersection of the lines.

(i) $2x^2 - 7xy + 3y^2 + 5x - 5y + 2$
(ii) $y^2 - 4xy + 4y - 4x + 3$
(iii) $2x^2 + xy - y^2 - 2x + y$.

5.1.2 Write out a proof of Lemma 5.3 following that of Lemma 5.2.

5.1.3 In each of the following cases determine the affine types of the given conic in $\mathbb{R}^2$, and write down an affine equivalence giving the normal form.

(i) $x^2 - 2xy + y^2 + 2x + 3y + 2$
(ii) $3x^2 - 10xy + 3y^2 + 14x - 2y - 8$
(iii) $5x^2 + 6xy + 5y^2 - 4x + 4y - 4$.

5.1.4 In each of the following cases determine the affine typc of the given conic in $\mathbb{C}^2$, and write down an affine equivalence giving the normal form. In each case where the conic reduces to two lines, find the equations of the lines.

(i) $x^2 + xy - 2y^2 + x - y$
(ii) $x^2 - 4xy + 3y^2 + 2x$
(iii) $x^2 - 2xy + y^2 + 2x - 4y + 3$
(iv) $9x^2 - 24xy + 16y^2 + 3x - 4y - 2$.

5.1.5 For each real number t let f_t be the conic in $\mathbb{R}^2$ defined by the formula $f_t(x, y) = x^2 + t(t+1)y^2 - 2txy + 2x + 2$. Determine the affine types of the f_t for all t.

5.1.6 For each real number t let f_t be the conic in $\mathbb{R}^2$ defined by the formula $f_t = 2y(x-1) + 2t(x-y)$. Determine the affine types of f_t for all t.

5.1.7 For each real number t let f_t be the conic in $\mathbb{R}^2$ which is defined by the formula $f_t(x, y) = x^2 - 2x - 2y - 1 + t(x+y)^2$. Determine the affine type of f_t for all t.

5.2 The Delta Invariants

Although we have lists of affine conics in both the real and the complex cases, we have as yet no way of distinguishing the various types on the lists. This section is devoted to the question of achieving such distinctions using two 'delta invariants' δ and Δ. The starting point is to note that the general conic

$$ax^2 + 2hxy + by^2 + 2gx + 2fy + c \qquad (\star)$$

can be written advantageously in matrix form. The symmetric 3×3 matrix

$$A = \begin{pmatrix} a & h & g \\ h & b & f \\ g & f & c \end{pmatrix}$$

is called the *matrix* of ($\star$). Write $v = (x, y, 1)$: then ($\star$) can be written vAv^T, where v^T denotes the transpose of the row vector v. We define $\Delta = \det A$, $\delta = \det B$ where B denotes the leading 2×2 submatrix of A; Δ is a somewhat complex expression in a, b, c, f, g, h, whilst δ is given by the simpler expression $\delta = ab - h^2$. In these circumstances it is convenient to write an affine mapping in a compatible matrix form, namely $v^T = PV^T$, where $V = (X, Y, 1)$ and P is a matrix of the following form with $\det P = ps - rq \neq 0$.

$$P = \begin{pmatrix} p & q & u \\ r & s & v \\ 0 & 0 & 1 \end{pmatrix}.$$

Lemma 5.4 *The effect of an affine transformation on a conic is to multiply δ, Δ by non-zero scalars. Thus the conditions $\delta = 0$, $\Delta = 0$ are invariant under affine transformations.*

Table 5.3. *Invariants for complex affine conics*

normal form	name	Δ	δ
$y^2 = x$	parabola	$\Delta \neq 0$	$\delta = 0$
$y^2 = -x^2 + 1$	general conic	$\Delta \neq 0$	$\delta \neq 0$
$y^2 = x^2$	line-pair	$\Delta = 0$	$\delta \neq 0$
$y^2 = 1$	parallel lines	$\Delta = 0$	$\delta = 0$
$y^2 = 0$	repeated line	$\Delta = 0$	$\delta = 0$

Proof Let A' be the matrix of ($\star$) in the X,Y coordinates. Under the change of coordinates, the conic becomes $vAv^T = (VP^T)A(VP^T)^T = VP^TAPV^T = VA'V^T$ where $A' = P^TAP$. Then $\det A' = \det P^TAP = \det P^T \det A \det P = (\det P)^2 \det A$; thus $\det A$, $\det A'$ differ by a non-zero scalar. That shows that under an affine change of coordinates Δ is multiplied by a non-zero scalar $(\det P)^2$. Now let B, B' be the leading 2×2 submatrices of A, A'; and let Q be the leading 2×2 submatrix of P, so $\det Q = ps - rq \neq 0$. Then by straight calculation $B' = Q^TBQ$, so $\det B' = \det Q^T \det B \det Q = (\det Q)^2 \det B$; thus $\det B$, $\det B'$ differ by a non-zero scalar. That shows that under an affine change of coordinates δ is likewise multiplied by a non-zero scalar $(\det Q)^2$. Under scalar multiplication, the conic (and hence the matrices A and B) are multiplied by a non-zero scalar λ; that multiplies the determinants δ and Δ by the non-zero scalars λ^2, λ^3. □

The δ- and Δ-invariants help us to distinguish the five types of complex affine conics. Since the conditions $\delta = 0$, $\Delta = 0$ are invariant under affine transformations, it suffices to consider normal forms. The reader is left to check that the invariants for the normal forms are given by Table 5.3. The only types which fail to be distinguished by the invariants are the parallel lines and repeated lines; however these are automatically distinguished by the multiplicities of the components. Incidentally, it follows from the table that *a conic in* $\mathbb{C}^2$ *is reducible if and only if* $\Delta = 0$.

Note the particular case of Lemma 5.4 when $\mathbb{K}$ is the real field. In that case, affine changes of coordinates and scalar multiplication multiply δ by *positive* scalars. Thus *over the real field affine equivalent conics have invariants* δ *with the same sign*. However the sign of Δ is *not* invariant over the real field; multiplication of the equation by -1 changes the sign of Δ. The invariants for the normal forms of real affine conics are given by Table 5.4. Looking at the table, we see that few types fail to

Table 5.4. *Invariants for real affine conics*

normal form	name	Δ	δ
$y^2 = x$	parabola	$\Delta \neq 0$	$\delta = 0$
$y^2 = -x^2 + 1$	real ellipse	$\Delta \neq 0$	$\delta > 0$
$y^2 = -x^2 - 1$	imaginary ellipse	$\Delta \neq 0$	$\delta > 0$
$y^2 = x^2 + 1$	hyperbola	$\Delta \neq 0$	$\delta < 0$
$y^2 = x^2$	real line-pair	$\Delta = 0$	$\delta < 0$
$y^2 = -x^2$	imaginary line-pair	$\Delta = 0$	$\delta > 0$
$y^2 = 1$	real parallel lines	$\Delta = 0$	$\delta = 0$
$y^2 = -1$	imaginary parallel lines	$\Delta = 0$	$\delta = 0$
$y^2 = 0$	repeated line	$\Delta = 0$	$\delta = 0$

be distinguished by the invariants. The real and imaginary ellipses (and likewise the real and imaginary parallel lines) are distinguished by the fact that the zero set of the latter is empty, whilst that of the former is non-empty. And (as in the complex case) the repeated line is distinguished from all other normal forms by the fact that it has a component of multiplicity 2. Note that in the real case, $\Delta = 0$ is no longer a necessary and sufficient condition for a conic to be reducible. For instance, $\Delta = 0$ for the imaginary parallel lines $y^2 = -1$, but the conic does not reduce over the real field.

Example 5.2 Consider the conic $x^2 - 2xy + 5y^2 + 2x - 10y + 1$. The reader will readily check that $\delta = 4 > 0$, $\Delta = 24 \neq 0$ so by Table 5.4 the conic is a real or an imaginary ellipse. In fact it is a real ellipse because it has real points. One way of seeing that is to intersect it with the real line $y = 0$ to get $x^2 + 2x + 1 = 0$, i.e. $(x + 1)^2 = 0$, yielding $x = -1$. Thus $(-1, 0)$ is a real point on the conic.

The delta invariants relate to the concept of a centre, discussed in Section 4.4. We saw there that $\delta \neq 0$ is precisely the condition for the conic to have a unique centre; when $\delta = 0$ we either have no centre, or a line of centres. You get a better idea of the geometric meaning of the Δ-invariant from the following result.

Lemma 5.5 *A sufficient condition for $\Delta = 0$ is that the conic $(\star)$ has a centre lying on the conic. The condition is also necessary, provided $\delta \neq 0$, i.e. the centre is unique.*

Proof Let (α, β) be a centre, so equations (5.1) and (5.2) are satisfied automatically. We claim first that the centre lies on the conic if and only if the following system of linear equations has a solution (α, β, γ) with $\gamma = 1$.

$$a\alpha + h\beta + g\gamma = 0 \quad (5.1)$$

$$h\alpha + b\beta + f\gamma = 0 \quad (5.2)$$

$$g\alpha + f\beta + c\gamma = 0. \quad (5.3)$$

Suppose the centre lies on the conic, so $x = \alpha$, $y = \beta$ satisfies the equation ($\star$). Multiplying (5.1) by α, (5.2) by β, and subtracting the two resulting equations from ($\star$) with $x = \alpha$, $y = \beta$ we get (5.3). Conversely, suppose (5.3) is satisfied. Multiplying (5.1) by α, (5.2) by β, and adding (5.3) we get ($\star$) with $x = \alpha$, $y = \beta$, so the centre (α, β) lies on the conic ($\star$). By linear algebra, the system has a non-trivial solution if and only if the determinant of the matrix of coefficients vanishes, i.e. if and only if $\Delta = 0$. That establishes the first part of the proposition. For the second part, observe that if $\delta \neq 0$ then any non-trivial solution of the system must have $\gamma \neq 0$, and then multiplying through by $1/\gamma$ we get a solution with $\gamma = 1$. □

Exercises

5.2.1 Determine the affine type of the conic $5x^2+6xy+5y^2-4x+4y-4$ in $\mathbb{R}^2$ by calculating the delta invariants.

5.2.2 Calculate the delta invariants for the prenormal form $\psi = y^2 - \alpha x^2 - 2\beta x - \gamma$ in $\mathbb{C}^2$. Show that ψ is a parabola if and only if $\alpha = 0$ and $\beta \neq 0$, and is a repeated line if and only if $\alpha = 0$, $\beta = 0$ and $\gamma = 0$.

5.2.3 For each real number t a conic f_t in $\mathbb{C}^2$ is defined by the formula

$$f_t = x^2 + t(t+1)y^2 + 2 - 2txy + 2x.$$

Show that all the f_t are central, and that for $t \neq 0$ the centres are collinear. Express the delta invariants in terms of t, and hence determine the affine type of f_t for all t.

5.2.4 For each real number t a conic in $\mathbb{R}^2$ is defined by the formula

$$f_t = 2y(x-1) + 2t(x-y).$$

Show that all the f_t are central, and that the centres are collinear. Express the delta invariants in terms of t, and hence determine the affine type of f_t for all t.

5.2.5 For each real number t a conic in $\mathbb{R}^2$ is defined by the formula

$$f_t = x^2 - 2x - 2y - 1 + t(x+y)^2.$$

Express the delta invariants in terms of t. By considering the intersections of f_t with the line $x = 1$, show that f_t has a real point, for any choice of t. Determine the affine type of f_t for all t.

5.2.6 For each real number t, a conic in $\mathbb{R}^2$ is defined by the formula

$$f_t = t(x^2 + y^2) - (x-1)^2.$$

Show that f_t contains no real points for $t < 0$. Also, by considering the intersections of f_t with appropriate lines, show that f_t has at least one real point for all $t \geq 0$. Express the delta invariants in terms of t. Determine the affine type of f_t for all t. Further, show that all the centres of the f_t are collinear.

5.2.7 In classical Greek mathematics, conics were obtained as the sections of the cone $z^2 = x^2 + y^2$ in $\mathbb{R}^3$ with a plane $z = \alpha x + \beta y + \gamma$. Obtain a defining equation by eliminating z from these two relations, and show that the delta invariants are given by $\delta = 1 - \alpha^2 - \beta^2$, $\Delta = \gamma^2$. Show that six of the nine real affine types of conics can be obtained in this way, but that three cannot.

5.2.8 In Example 5.2.8, it was shown that the trajectory of a general point $P = (X, Y)$ under the planar motion generated by the double slider is the conic

$$((d-X)x - Yy)^2 + (-Yx + Xy)^2 = F(X,Y)^2$$

where $d > 0$, and $F(X,Y) = X^2 + Y^2 - dX$. Calculate the invariants δ, Δ in terms of X, Y and d and hence verify that the conic is an ellipse if and only if $F(X,Y) \neq 0$.

5.3 Uniqueness of Equations

In Example 1.1, we spelled out an important property of affine lines, namely that their equations are determined by their zero sets. It is revealing to adopt the same simple minded approach to this question for conics. Example 1.1 depended on the fact that the three coefficients in the equation of a line are determined (up to a scalar multiple) by two linearly independent conditions. The general principle is that N linearly independent conditions on $(N+1)$ unknowns determine those unknowns, up to a non-zero scalar multiple. Let us try to apply this to conics. A general conic in $\mathbb{K}^2$ has the form ($\star$) with six coefficients a, b, c, f, g, h, so we should be looking for five linearly independent conditions. In principle we expect such conditions to result from the requirement that ($\star$) passes through five points $p_k = (x_k, y_k)$. That yields five linear conditions on a, b, c, f, g, h.

$$\begin{array}{rcl} ax_1^2 + 2hx_1y_1 + by_1^2 + 2gx_1 + 2fy_1 + c & = & 0 \\ \vdots & \vdots & \vdots \\ ax_5^2 + 2hx_5y_5 + by_5^2 + 2gx_5 + 2fy_5 + c & = & 0 \end{array} \tag{5.4}$$

These equations have at least one non-trivial solution, establishing a useful generality, namely that *through any five points in the plane there passes at least one conic*. The serious question is whether these conditions are linearly independent.

Lemma 5.6 *Suppose that the zero set of an irreducible conic in* $\mathbb{K}^2$ *contains at least five distinct points* p_k. *Then the linear equations (5.4) are independent, and hence any defining polynomial for the conic is determined up to a non-zero scalar multiple.*

Proof Suppose the equations are dependent so that at least one is a non-trivial linear combination of the others. By symmetry we can suppose that the last equation is a non-trivial linear combination of the first four. That means that *any* conic passing through p_1, p_2, p_3, p_4 automatically passes through p_5. For $i \neq j$ write l_{ij} for the line joining p_i, p_j. Then the line-pair comprising the lines l_{13}, l_{24} passes through p_5, so either p_1, p_3, p_5 are collinear, or p_2, p_4, p_5 are collinear. Either way we contradict the fact that a line meets an irreducible conic in at most two points. □

In particular, irreducible conics in $\mathbb{C}^2$ have unique equations; we know from Lemma 2.1 that their zero sets are infinite, so automatically contain five distinct points. Likewise, conics in $\mathbb{R}^2$ of parabola, real ellipse or hyperbola type have unique equations since their zero sets are infinite. However, the *hypothesis* of Lemma 5.6 fails for imaginary ellipses, as does the *conclusion*. For instance the conics $x^2 + y^2 + 1$, $x^2 + y^2 + 2$ in $\mathbb{R}^2$ have the same zero set (namely the empty set) but do not differ by a scalar multiple.

Exercises

5.3.1 Show that any *reducible* conic in $\mathbb{K}^2$ is determined by its zero set, i.e. that any two polynomials having the same zero set differ by a non-zero scalar. (Consider separately the cases of line-pairs and repeated lines.) In view of this happy accident *any* conic in $\mathbb{C}^2$ has a unique equation.

5.3.2 Clearly, the hypothesis of Lemma 5.6 can be weakened by assuming only that the zero set contains five (not necessarily distinct) points, no three of which are collinear. Show that it suffices to assume a yet weaker hypothesis, that the zero set contains at least five (not necessarily distinct) points, no four of which lie on a line.

6

Singularities of Affine Curves

When turning the pages of this book, you will see numerous computer generated pictures of algebraic curves. What should strike you about many of the pictures is that curves can possess points which in a visual sense are 'singular'; for instance, points where the curve crosses itself, or has sharp 'cusps', or even isolated points. Understanding these 'singular' points is one of the more important objectives of the subject – the further one delves into the study of curves, the more important the 'singularities' become.

The starting point for such a study is to look in detail at the way in which lines intersect a curve f, developing the germ of an idea implicit in the proof of Lemma 4.4, namely that the intersections correspond to the zeros of a certain polynomial (the 'intersection polynomial') and can be counted properly by their 'intersection numbers'. In the following section, we will specialize the situation by considering only those lines which pass through a given point p on f; that will enable us to associate to p a positive integer known as the 'multiplicity' of p on f. It will turn out that there are at most finitely many points on an irreducible curve for which the 'multiplicity' is ≥ 2; these are the so-called 'singular' points of f.

6.1 Intersection Numbers

Let f be a curve of degree d in $\mathbb{K}^2$, and let l be a line which is not a component of f. Then by Lemma 4.4, f and l meet in $\leq d$ points. Let p be a point on l, and suppose (for a moment) that p lies on f. Generally, if we disturb l slightly we expect to have just one intersection close to p. Take for instance f to be the parabola $f(x, y) = y - x^2$ in $\mathbb{R}^2$, l to be the line $l = y$ and $p = (0, 0)$, then for arbitrarily small t the line $y - tx$ will have *two* intersections with f close to p; as t becomes small we imagine

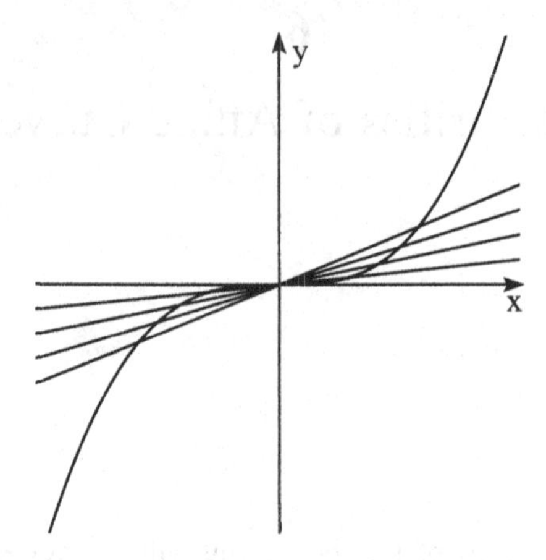

Fig. 6.1. Coalescence of intersections

the two intersections coalescing, and in the limit we think of f having *two* 'coincident' intersections with l at p. However, that is not always the pattern. If we replace f by the cubic $f(x,y) = y - x^3$, then for arbitrarily small positive t the line $y - tx$ will have *three* intersections with f close to p, and in the limit we think of f as having *three* 'coincident' intersections with l at p. The mental picture is illustrated in Figure 6.1.

An algebraic formalization of this geometric intuition can be obtained as follows. Let a, b be any two distinct points on l. Then l is parametrized as $(1-t)a + tb$ with $t \in \mathbb{K}$, and the intersections of f and l are given by $\phi(t) = 0$ where $\phi(t)$ is the *intersection polynomial* defined by $\phi(t) = f((1-t)a + tb)$. More explicitly, if $a = (a_1, a_2)$ and $b = (b_1, b_2)$ then the parametrization is $x(t) = (1-t)a_1 + tb_1$, $y(t) = (1-t)a_2 + tb_2$, and $\phi(t) = f(x(t), y(t))$. Note that $\phi(t)$ has degree $\leq d$; the examples below show that the degree can be $< d$. Let p be a point on l with parameter t_0. We define the *intersection number* $I(p, f, l)$ to be the multiplicity of t_0 as a root of the polynomial equation $\phi(t) = 0$, i.e. it is the highest power of $(t - t_0)$ in the factorization of $\phi(t)$.

Suppose l is not a component of f, then the intersection number $I(p, f, l)$ is an integer ≥ 0, and is $= 0$ if and only if p is not a point of intersection of f, l. (When l is a component of f, and p is a point on l, it is a convention that $I(p, f, l) = \infty$.) The *total number* of intersections of f with l is the sum of the intersection numbers $I(p, f, l)$ with p on l. As the sum of the multiplicities of the roots of $\phi(t)$ is $\leq d$ we obtain a useful general principle: *the total number of intersections of a curve f of* *degree d with a line l is $\leq d$.* Another way of expressing this is that the number of intersections 'counted properly' is $\leq d$.

The next step in our development is to establish that $I(p, f, l)$ does not depend on the chosen parametrization of the line l. For this we isolate a useful little lemma, requiring a preliminary definition. By an affine *change of parameter* we mean a mapping $u : \mathbb{K} \to \mathbb{K}$ of the form $u(t) = at + b$ with $a \neq 0$. We now have

Lemma 6.1 *Suppose that two polynomials $\phi(t)$, $\psi(u)$ in one variable are related by an affine change of parameter, i.e. $\phi(t) = \psi(u(t))$ for some affine map $u(t)$. Then t_0 is a zero of multiplicity m of $\phi(t)$ if and only if $u(t_0)$ is a zero of multiplicity m of $\psi(u)$.*

Proof Observe (using the Factor Theorem) that $t - t_0$ is a factor of $\phi(t)$ if and only if t_0 is a zero of $\phi(t)$, if and only if $u(t_0)$ is a zero of $\psi(u)$, if and only if $u - u(t_0)$ is a factor of $\psi(u)$. However, by Lemma 4.6 corresponding factors $t - t_0$, $u - u(t_0)$ have the same multiplicity, as required. □

Lemma 6.2 *The intersection number $I(p, f, l)$ depends only on p, f, l; it does not depend on the choice of parametrization of l.*

Proof Consider two parametrizations of l, the first via two distinct points a, b with parameter t, yielding the intersection polynomial $\phi(t) = f(ta + (1-t)b)$; and the second via two distinct points c, d with parameter u, yielding the intersection polynomial $\psi(u) = f(uc + (1-u)d)$. To establish a relation between ϕ, ψ write $a = \alpha c + (1-\alpha)d$, $b = \beta c + (1-\beta)d$: then, by direct calculation

$$\phi(t) = f(ta + (1-t)b) = f(u(t)c + (1-u(t))d) = \psi(u(t))$$

where $u(t) = (\alpha - \beta)t + \beta$. Thus $\phi(t)$, $\psi(u)$ are related by the affine change of parameter $u(t)$. Then Lemma 6.1 tells us that t_0 is a zero of $\phi(t)$ of multiplicity m if and only if $u(t_0)$ is a zero of $\psi(u)$ of multiplicity m. The result follows from the definition of $I(p, f, l)$. □

Example 6.1 The curve $f = xy - 1$ is a conic in $\mathbb{K}^2$. Let l be the x-axis, parametrized as $x(t) = t$, $y(t) = 0$. Then $\phi(t) = -1$ has degree $0 < 2$. There are *no* intersections, and $I(p, f, l) = 0$ for every point p on l.

Example 6.2 The conic $f = y - x^2$ is a parabola in $\mathbb{R}^2$. Consider the line $y = \lambda x$ through $p = (0, 0)$, parametrized as $x = t$, $y = \lambda t$. Then

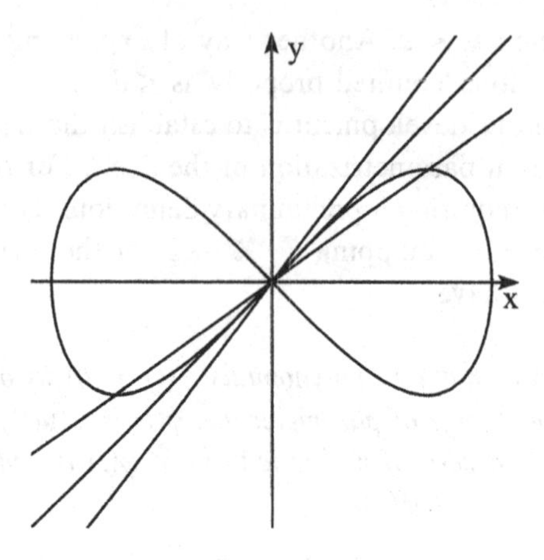

Fig. 6.2. Lines intersecting the eight-curve

$\phi(t) = \lambda t - t^2 = t(\lambda - t)$ with the zero $t = 0$ corresponding to an intersection at p, and the zero $t = \lambda$ corresponding to an intersection elsewhere. For $\lambda \neq 0$ the zero $t = 0$ has multiplicity 1 and $I(p, f, l) = 1$, but for $\lambda = 0$ the zero $t = 0$ has multiplicity 2 and $I(p, f, l) = 2$. Thus there is an *exceptional* line through p, namely $y = 0$, which meets the parabola twice at that point.

Example 6.3 The quartic $f = y^2 - x^2 + x^4$ is known as the *eight-curve*. Consider the line $y = \lambda x$ through $p = (0,0)$, parametrized as $x = t, y = \lambda t$. (Figure 6.2.) Then $\phi(t) = \lambda^2 t^2 - t^2 + t^4 = t^2(t^2 + \lambda^2 - 1)$. The zeros of $\phi(t)$ depend on λ. For $-1 < \lambda < 1$, there is a zero $t = 0$ of multiplicity 2, and two more of multiplicity 1; for instance, when $\lambda = 0$ we have roots $t = 1$ and $t = -1$ of multiplicity 1, and $t = 0$ of multiplicity 2, so l meets the curve at $(-1,0)$, $(0,0)$, $(1,0)$ with respective intersection numbers 1, 2, 1. For $\lambda < -1$ or $\lambda > 1$, there is just one root $t = 0$ of multiplicity 2, so just one intersection at p with intersection number $I(p, f, l) = 2$. Finally, for $\lambda = \pm 1$, there is just one root $t = 0$ of multiplicity 4, so just one intersection p with intersection number $I(p, f, l) = 4$. The moral of this example is that *most* lines through p meet the curve twice at p, but there are two *exceptional* lines which meet the curve more than twice at p.

Lemma 6.3 *Intersection numbers are invariant under affine maps; if we apply an affine map α to a point p, a line l through p, and a curve f*

to obtain a point p', a line l' through p', and a curve f' then we have $I(p, f, l) = I(p', f', l')$.

Proof Let a, b be distinct points on l; then l is parametrized as $(1-t)a+tb$, and l' is parametrized as $(1-t)\alpha(a)+t\alpha(b)$. The intersection polynomial associated to f is $\phi(t) = f((1-t)a+tb)$, and the intersection polynomial associated to $f' = f \circ \alpha^{-1}$ is

$$\begin{aligned}
\phi'(t) &= f'\{(1-t)\alpha(a)+t\alpha(b)\} \\
&= f\left(\alpha^{-1}\{(1-t)\alpha(a)+t\alpha(b)\}\right) \\
&= f\left((1-t)\alpha^{-1}(\alpha(a))+t\alpha^{-1}(\alpha(b))\right) \\
&= f((1-t)a+tb) = \phi(t).
\end{aligned}$$

Now suppose p has parameter t on l. Then by Example 1.4, p' has parameter t on l' and the intersection numbers $I(p, f, l)$, $I(p', f', l')$ are the respective multiplicities of t as a zero of the intersection polynomials $\phi(t)$, $\phi'(t)$ and hence equal. □

Putting the bits together we obtain the following picture of how a curve f of degree d in $\mathbb{K}^2$ meets a line l which is not a component of f. There are $\leq d$ distinct points of intersection $p_1, \dots, p_s$ to which can be associated intersection numbers $i_1, \dots, i_s$, where $i_k = I(p_k, f, l)$, with $i_1 + \cdots + i_s \leq d$. The sum $i_1 + \cdots + i_s \leq d$ is the *intersection pattern* of l with f, and is invariant under affine mappings.

Example 6.4 The intersection pattern of the ellipse $x^2+y^2 = 1$ in $\mathbb{R}^2$ with a line l meeting the ellipse is either $1+1$ or 2. Let $p = (a, b)$, $q = (c, d)$ be distinct points on l, giving the parametrization $x = (1-t)a+tc$, $y = (1-t)b+td$. Substituting for x, y in the normal form we obtain a real quadratic equation $At^2+Bt+C = 0$ where $A = (a-c)^2+(b-d)^2$, $B = a(c-2)+b(d-2)$, $C = a^2+b^2-1$. Clearly $A \neq 0$ (else the points p, q coincide) so either there are two real roots of multiplicity 1, or there is a repeated real root of multiplicity 2, i.e. l meets the ellipse twice with intersection number 1, or just once with intersection number 2. Note that the argument fails over the complex field, for instance, the line $y = ix+1$ meets the ellipse at just one point $(0, 1)$ with multiplicity 1.

Affine curves can be distinguished by their intersection patterns with lines. For instance the normal form $y^2 = x$ for the parabola in $\mathbb{R}^2$ has the property that there exists a line (namely $y = 0$) which meets it just once, with intersection number 1; however, the above example shows that the

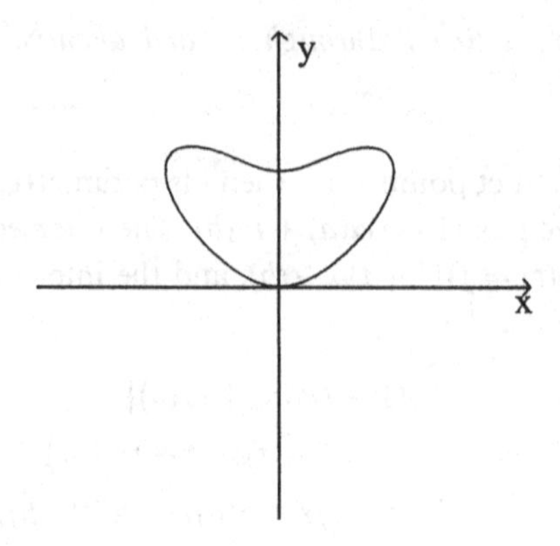

Fig. 6.3. The boomerang

normal form $x^2 + y^2 = 1$ for the ellipse fails to have that property. It follows from the affine classification of real conics in Table 5.1 that in $\mathbb{R}^2$ a parabola cannot be affine equivalent to an ellipse.

Example 6.5 The quartic curve in $\mathbb{R}^2$ defined by $f = y(x^2+y^2)-x^4-y^4$ is known as the *boomerang*. (Figure 6.3.) The picture suggests that there is a horizontal line which meets the curve just twice, with intersection pattern $2+2$. To verify this, look at the intersections with general 'horizontal' lines $y = c$. Parametrize the line as $x = t$, $y = c$. The intersection function is obtained by setting $x = t$, $y = c$ in the equation $f(x, y) = 0$, giving a polynomial equation $\phi(t) = f(t, c) = 0$ of degree 4 in t. In principle there are four (possibly complex) values of t with $\phi(t) = 0$, each *real* solution corresponding to an intersection of the line $y = c$ with the curve. And the intersection number corresponding to such a real root t is ≥ 2 provided t is a *repeated* real root of $\phi(t) = 0$. We are looking for a value of c with the property that $\phi(t) = 0$ has two distinct repeated real roots. The key observation is that $\phi(t) = 0$ is a *quadratic* in $T = t^2$, namely $-T^2 + cT - c^3(c-1) = 0$, so has *either* two distinct roots, *or* a repeated real root. Clearly, in the former case the quartic in t cannot have two repeated real roots, *so we require the quadratic in T to have a repeated root*. Recall that the condition for a quadratic to have a repeated root is that its discriminant vanishes. By calculation, the discriminant is $c^2(1+4c-4c^2)$, which vanishes if and only if $c = 0$ or $c = \frac{1}{2}(1 \pm \sqrt{2})$. Now consider cases. The line $y = c = 0$ meets the curve when $t^4 = 0$, so just once with intersection number 4. The lines $y = c = \frac{1}{2}(1 \pm \sqrt{2})$ meet the curve when $t^2 = c/2$. In the '+' case $c > 0$ and we obtain just two values of t, hence just two points of intersection, with corresponding intersection number 2, hence the required intersection pattern $2+2$. In the '$-$' case the equation $t^2 = c/2$ has no solutions, and there are no intersections, as the picture confirms.

It is worth pointing out that the concept of the intersection number of a line l with a curve f in $\mathbb{K}^2$ depended only on the fact that lines always have parametrizations $x = x(t)$, $y = y(t)$. The concept can be extended from lines l to any curve with this property, by defining the intersection function $\phi(t) = f(x(t), y(t))$ and proceeding as above.

Example 6.6 The parabola $y = x^2$ in $\mathbb{R}^2$ has the parametrization $x = t$, $y = t^2$. To investigate its intersections with the circle $f = x^2 + y^2 - 2y = 0$ of radius 1 centred at $(0, 1)$ we form the intersection function $\phi(t) = t^4 - t^2 = t^2(t-1)(t+1)$ having zeros $t = 0$, $t = 1$, $t = -1$ of respective multiplicities 2, 1, 1. There are therefore three intersections, at the points $(0, 0)$, $(1, 1)$, $(-1, 1)$ to which can be associated 'intersection numbers' 2, 1, 1.

It is tempting to think that this idea can be used to define intersection numbers for any two curves in $\mathbb{K}^2$. Unfortunately it founders on a very basic fact, namely that very few curves have 'global' parametrizations $x = x(t)$, $y = y(t)$. We will say rather more about this topic in Chapter 8. The introduction of 'intersection numbers' for two general curves requires a new set of ideas, explained in Chapter 14.

6.2 Multiplicity of a Point on a Curve

Now specialize the above by taking $p = (a, b)$ to be a fixed point on the curve f, and consider the intersections of f with the 'pencil' of all lines through p. (We will give a formal definition of this term in Section 16.2.) In this situation, $I(p, f, l) \geq 1$ automatically. The *multiplicity* of p on f is the minimal value m of the intersection number $I(p, f, l)$ taken over all lines l through p, i.e. $I(p, f, l) \geq m$ for *every* line l through p, and there exist lines l for which $I(p, f, l) = m$. Since intersection numbers are affine invariants, the multiplicity of a point on a curve is likewise an affine invariant. Points of multiplicity 1, 2, 3, 4, ... are said to be *simple*, *double*, *triple*, *quadruple*, ... points of f. A crude mental picture for a point p of multiplicity m is that there are m 'branches' of the curve self-intersecting at p. (But beware; as we will see, that is not necessarily the case.) For

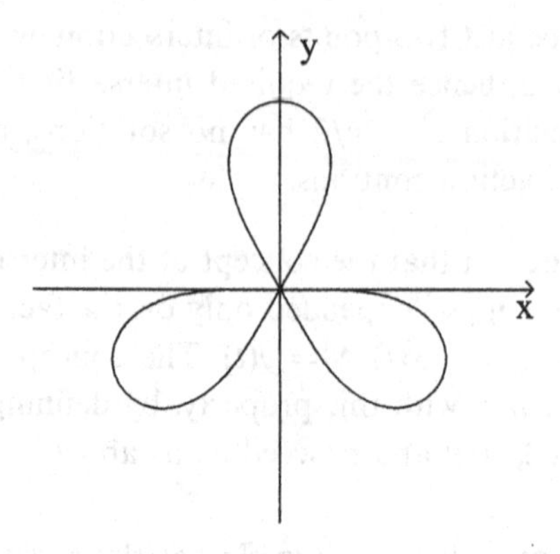

Fig. 6.4. Triple point on the three-leaved clover

instance the three-leaved clover of Figure 6.4 has a point of multiplicity 3 at the origin.

The next result provides a computational technique for determining the multiplicity of a point on a curve. Recall that the 'lowest order terms' (LOT) in a polynomial were defined in Section 3.4.

Lemma 6.4 *Let $p = (a, b)$ be a point on a curve f in $\mathbb{K}^2$. Set $g(x, y) = f(a + x, b + y)$. The multiplicity of p on f is the degree of the LOT in g. In particular, the multiplicity of the origin $(0, 0)$ on f is the degree of the LOT in f itself.*

Proof Write $g = G_m + G_{m+1} + \cdots + G_d$ where each G_k is a form of degree k, and $G_m \neq 0$. Let (X, Y) be a non-zero vector in $\mathbb{K}^2$, and let $l = l_{X,Y}$ be the line joining the points (a, b), $(a+X, b+Y)$, thus l is given parametrically as $x(t) = a+tX$, $y(t) = b+tY$. The associated intersection polynomial is then

$$\begin{aligned}\phi_{X,Y}(t) &= f(a+tX, b+tY)\\ &= g(tX, tY)\\ &= G_m(tX, tY) + G_{m+1}(tX, tY) + \cdots\\ &= t^m G_m(X, Y) + t^{m+1} G_{m+1}(X, Y) + \ldots.\end{aligned}$$

Clearly, t^m is a factor of $\phi_{X,Y}(t)$ for all choices of X, Y. However, there exist choices of X, Y for which $G_m(X, Y) \neq 0$, and hence for which t^{m+1} is not a factor of $\phi_{X,Y}(t)$. (Recall that a binary form has only finitely many zeros.) It follows that p is a point of multiplicity m on f. □

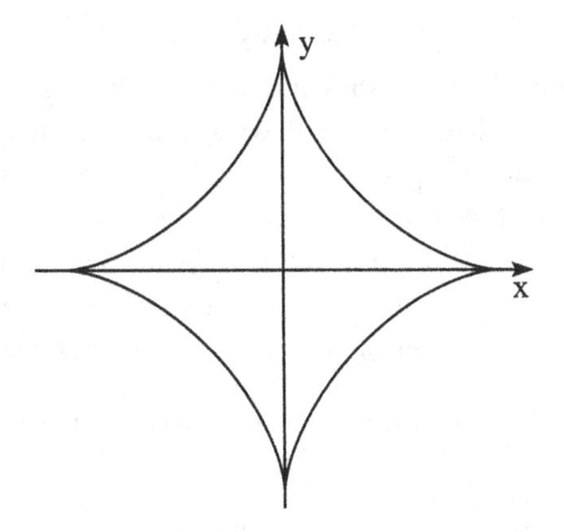

Fig. 6.5. The astroid

Example 6.7 The *astroid* is the curve in $\mathbb{R}^2$ given parametrically by $x = \sin^3 t$, $y = \cos^3 t$. By substitution, one can check that every point on the astroid lies on the sextic curve $f = (1 - x^2 - y^2)^3 - 27x^2y^2$. (In fact the zero set of f coincides with the image of the parametrized curve, but that is harder to check.) The illustration in Figure 6.5 suggests that four points on the curve, namely $(1,0)$, $(0,1)$, $(-1,0)$, $(0,-1)$ are in some sense exceptional. For instance the point $(1,0)$ has multiplicity 2; using Lemma 6.4 with $a = 1$, $b = 0$ we see that

$$g(x,y) = f(x+1,y) = -(x^2+2x+y^2)^3 - 27(x+1)^2y^2 = -27y^2 + \text{ HOT }.$$

The criterion for a point to be of given multiplicity m on a curve can also be phrased usefully in terms of partial derivatives.

Lemma 6.5 *Let $p = (a,b)$ be a point on a curve f in $\mathbb{K}^2$. Then p has multiplicity m on f if and only if all the partial derivatives of f of order $< m$ vanish at p, and at least one partial derivative of order m is non-zero at p.*

Proof Set $g(x,y) = f(a+x, b+y)$, and write $g = G_0 + \cdots + G_d$ where each G_k is a form of degree k. Then by Lemma 6.4 the point p has multiplicity m on f if and only if $G_0 = \ldots = G_{m-1} = 0$, $G_m \neq 0$. The Taylor expansion of $g(x,y)$ shows that these conditions hold if and only if all the partial derivatives of g of order $< m$ vanish at the origin, and at least one of order m is non-zero at the origin. The result follows since the partial derivatives of g at the origin coincide with those of f at p. □

Exercises

6.2.1 In Example 1.8 it was shown that the pedal curve of the standard parabola $y^2 - 4ax$ with $a > 0$, with respect to the pedal point $p = (\alpha, 0)$ satisfies the equation of the cubic $f = x(x-\alpha)^2 + y^2(a-\alpha+x)$ in $\mathbb{R}^2$. Show that p is a double point of f.

6.2.2 In Example 1.10 it was shown that the pedal curve of the unit circle $x^2 + y^2 = 1$ with respect to the pedal point $p = (\alpha, 0)$ satisfies the equation of the quartic in $\mathbb{R}^2$ defined by

$$f = \{x(x-\alpha) + y^2\}^2 - (x-\alpha)^2 - y^2.$$

Show that p is a double point of f.

6.2.3 In each of the following cases, find the multiplicity of the given point p on the curve f in $\mathbb{R}^2$.

(i) $p = (0, 1)$ $f = (x^2 - 1)^2 - y^2(3 - 2y)$
(ii) $p = (0, 1)$ $f = x^2(3y - 2x^2) - y^2(1 - y)^2$
(iii) $p = (0, 1)$ $f = x^2 y + xy(y - 1) + (y - 1)^2$
(iv) $p = (1, 1)$ $f = x^3 + y^3 - 3x^2 - 3y^2 + 3xy + 1$
(v) $p = (0, 4)$ $f = y^3(4 - y)^3 - 4x^4(x + 3)^2$.

6.3 Singular Points

Points of multiplicity ≥ 2 on a curve f are *singular* and play an important role in understanding the geometry of the curve. The curve f is *singular* when it has a singular point, otherwise it is *non-singular*. Since multiplicity is invariant under affine mappings, so too is the concept of a singular point. We will see in Chapter 14 that an irreducible curve f has at most finitely many singular points, thus the number of singular points, and the list of their multiplicities, is an affine invariant of f. Lemma 6.5 yields the following criterion for a point to be singular on a curve f.

Lemma 6.6 *A necessary and sufficient condition for a point $p = (a, b)$ to be singular on a curve f is that $f(p) = 0$, $f_x(p) = 0$, $f_y(p) = 0$.*

One finds the singular points of f by solving the equations displayed in the statement; there is no general technique beyond native ingenuity.

Example 6.8 The singular points of a line $f = ax + by + c$ are given by $0 = f$, $0 = f_x = a$, $0 = f_y = b$. These relations can only be satisfied when $a = b = 0$, which is impossible. Thus lines are non-singular.

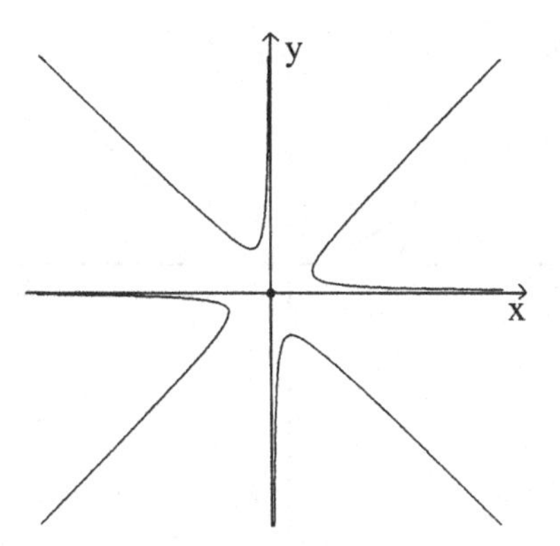

Fig. 6.6. The Maltese cross

Example 6.9 The singular points of a general conic f in $\mathbb{K}^2$ are given by $f(x,y)=0$, $f_x(x,y)=0$, $f_y(x,y)=0$. The latter relations say that (x,y) is a *centre* for f. *Thus a singular point of a conic is a centre which lies on the conic.* Over the complex field the only two normal forms having singular points are the line-pair (the singular point is the point of intersection) and the repeated line (every point on the line is singular); over the real field the outcome is the same, save that we have two types of line-pair, namely real and imaginary.

Example 6.10 For the eight-curve $f = y^2 - x^2 + x^4$, singular points require $0 = f$, $0 = f_x = 2x(2x^2-1)$, $0 = f_y = 2y$. Thus there is a *unique* singular point $(0,0)$, of multiplicity 2. It is no coincidence that this is the only point of the curve where the curve crosses itself.

Example 6.11 The *Maltese cross* is the quartic in $\mathbb{R}^2$ defined by $f = xy(x^2-y^2)-(x^2+y^2)$ and illustrated in Figure 6.6. Singular points are given by $0=f$, $0=f_x=3x^2y-y^3-2x$, $0=f_y=x^3-3xy^2-2y$. These relations look unpromising. The magic trick is to notice (using the relation $f=0$) that $0=xf_x+yf_y=4xy(x^2-y^2)-2(x^2+y^2)=2xy(x^2-y^2)$ so that $x=0$, $y=0$ or $x^2-y^2=0$. In each case it is clear that $(0,0)$ is the only solution of $f=0$, $f_x=0$, $f_y=0$. The magic trick is 'explained' in some measure by Exercise 10.3.9.

Example 6.12 Any curve in $\mathbb{R}^2$ can be viewed as a curve in $\mathbb{C}^2$; there may be singular points in $\mathbb{C}^2$ which are not in $\mathbb{R}^2$. Take for instance $f(x,y) = (1+x^2)^2 - xy^2$. Then $f_x = 4x(1+x^2) - y^2$, $f_y = -2xy$. For

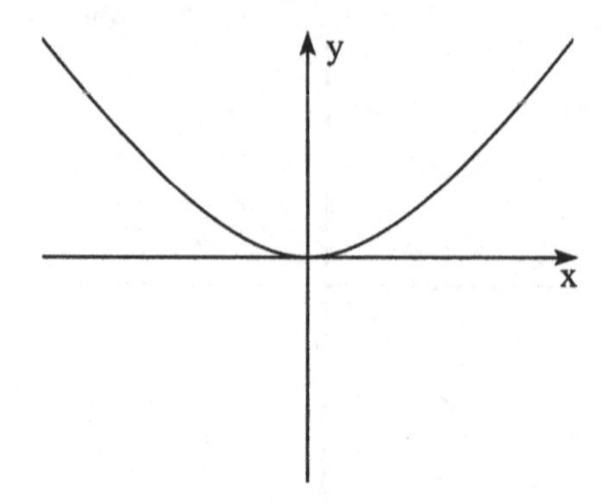

Fig. 6.7. An invisible singular point

singular points we have $x = 0$ or $y = 0$. $x = 0$ can't give a singular point as $f(0, y) = 1$. When $y = 0$, we have to satisfy $(1 + x^2)^2 = 0$, $4x(1 + x^2) = 0$. Over the real field these equations have no solution, and there are no singular points in $\mathbb{R}^2$. But over the complex field we have solutions $x = \pm i$, yielding singular points $(\pm i, 0)$ in $\mathbb{C}^2$.

Example 6.13 The reader is warned that *a singular point on a real curve may not be visible on the zero set*. An example is provided by the irreducible quartic $f = x^4 - x^2y - y^3$ in $\mathbb{R}^2$, having a singular point at the origin. The zero set of f is the image of the parametrized curve $x = t + t^3$, $y = t^2 + t^4$. (Exercise 8.1.12.) It follows that given x, the equation $f(x, y) = 0$ can be solved uniquely for y; indeed $t + t^3$ is a non-decreasing function of t with image the whole real line, so given, x we determine a unique t with $x = t + t^3$, hence a unique $y = t^2 + t^4$. And the tangent vector to the parametrization varies smoothly with t, so the zero set is perfectly 'smooth' at the origin, as can be seen in Figure 6.7.

One way in which singular points arise is as intersections of components of reducible curves.

Lemma 6.7 *Let p be a point of multiplicities m, n on the curves f, g in $\mathbb{K}^2$. Then p has multiplicity $m + n$ on the product $h = fg$. In particular, any intersection is singular on h.*

Proof Since multiplicities are invariant under affine mappings, it is no restriction to suppose that $p = (0, 0)$. By Lemma 6.4, we can write

$f = F_m +$ HOT, $g = G_n +$ HOT where F_m, G_n are non-zero forms of respective degrees m and n. Then $h = H_{m+n} +$ HOT, where $H_{m+n} = F_m G_n$ is a non-zero form of degree $m+n$. It follows immediately from Lemma 6.4 that p has multiplicity $m+n$ on h. □

For the reader with some knowledge of calculus in several variables we point out that the Implicit Function Theorem provides a description of a simple point $p = (a, b)$ on a real algebraic curve $f(x, y)$. By that result, there exist convergent power series $x(t)$, $y(t)$ defined on a neighbourhood of $t = 0$, with $a = x(0)$, $b = y(0)$ for which $f(x(t), y(t)) \equiv 0$ identically in t, and for which the parametric curve $z(t) = (x(t), y(t))$ is regular. In other words, in a neighbourhood of p, the set defined by $f(x, y) = 0$ is the image of a regular parametrized curve. Although this elegant description fails at singular points, the interested reader will find a natural generalization in more advanced texts using the idea of 'Puiseaux series'. Here is an application.

Example 6.14 This is one of the very few points in this text where we will make use of the Euclidean structure on the real affine plane $\mathbb{R}^2$. The *open disc* $B(p : \delta)$ of radius $\delta > 0$, centred at the point $p \in \mathbb{R}^2$ is defined to be

$$B(p : \delta) = \{q \in \mathbb{R}^2 : |p - q| < \delta\}$$

where $|p - q|$ denotes the usual Euclidean distance between the points p, q. A point p on a curve f in $\mathbb{R}^2$ is *isolated* when there exists a real number $\delta > 0$ such that p is the only point in $B(p : \delta)$ which lies on f. We claim that *an isolated point p on a curve f is necessarily singular.* Indeed, were p simple the intersection of the zero set of f with a sufficiently small disc centred at p would be the image of a regular parametrized curve, and (by the above description) p would fail to be simple.

Exercises

6.3.1 Find the singular points of the algebraic curve in $\mathbb{R}^2$ given by $f(x, y) = y^2 + 2x^2 - x^4 = 0$, and sketch the curve.

6.3.2 In each of the following cases, show that the curve $f(x, y)$ in $\mathbb{R}^2$ has just one singular point.

(i) $f(x, y) = x^3 + x^2 + y^2$

(ii) $f(x, y) = x^6 - x^2y^3 - y^5$

(iii) $f(x, y) = (x^2 + y^2)^2 - y(3x^2 - y^2)$.

6.3.3 In each of the following cases, show that the curve $f(x,y)$ in $\mathbb{R}^2$ has just one singular point.

(i) $f(x,y) = (x^2+y^2)^3 - 4x^2y^2$
(ii) $f(x,y) = x^2y^2 + 36x + 24y + 108$
(iii) $f(x,y) = y^3 + x^2 + 2xy - 6y^2 - 2x + 14y - 11$.

6.3.4 In each of the following cases, find all the singular points of the given algebraic curve $f(x,y)$ in $\mathbb{R}^2$, and show that they are double points.

(i) $f = (x^2-1)^2 - y^2(3+2y)$
(ii) $f = x^2(3y-2x^2) - y^2(1-y)^2$.

6.3.5 Let a, b be positive real numbers. The quartic curve f defined by $a^2y^2 - b^2x^2 = x^2y^2$ is known as the *bullet nose*. Show that the curve has exactly one singular point in $\mathbb{R}^2$, and no further singular points in $\mathbb{C}^2$.

6.3.6 Show that the astroid $f = (1-x^2-y^2)^3 - 27x^2y^2$ has exactly four singular points in $\mathbb{R}^2$, and exactly four more when regarded as a curve in $\mathbb{C}^2$.

6.3.7 Show that $p = (\alpha, 0)$ is the *only* singularity of the cubic curve $f = x(x-\alpha)^2 + y^2(a-\alpha+x)$ in $\mathbb{R}^2$. (f is the pedal curve of the standard parabola $y^2 - 4ax$ with respect to p.)

6.3.8 Show that $p = (\alpha, 0)$ is the *only* singularity of the quartic curve $f = \{x(x-\alpha)+y^2\}^2 - (x-\alpha)^2 - y^2$ in $\mathbb{R}^2$. (f is the pedal curve of the unit circle $x^2+y^2 = 1$ with respect to the pedal point p.)

6.3.9 This exercise relates to Example 2.9. Let $\phi(x)$ be a polynomial of degree $d \geq 3$ with coefficients in $\mathbb{K}$. Verify that every singular point of the affine curve $f = y^2 - \phi(x)$ in $\mathbb{K}^2$ must lie on the line $y = 0$. Show that the point $(a, 0)$ is a singular point of f if and only if a is a repeated root of $\phi(x)$. Use these results to show that the curve $f = y^2 - (5-x^2)(4x^4 - 20x^2 + 25)$ in $\mathbb{R}^2$ has exactly two singularities.

6.3.10 Let f be a curve of degree d, having at least m distinct singular points lying on a line l. Show that if $2m > d$ then l is a component of f.

7
Tangents to Affine Curves

In this chapter we introduce 'tangents' at a point p on an affine curve f via the intersection numbers of Chapter 6. Intuitively, a tangent at p is a line l through p which has higher contact with f than one would expect. The most general case (in some sense) is when p is a simple point on f, and there is a unique tangent line l given by an explicit formula; this provides the content of Section 7.2. Normally the tangent l at a simple point p has contact of order 2 with f at p, but when it is ≥ 3, we have very special points on f called 'flexes', which play a potentially important role in understanding the geometry of a curve. In the remainder of the chapter, we give a number of examples illustrating how to find the tangents to affine curves at singular points p.

7.1 Generalities about Tangents

Let p be a point of multiplicity m on an algebraic curve f in $\mathbb{K}^2$. Then automatically $I(p,f,l) \geq m$ for every line l through p. A *tangent line* (or just *tangent*) to f at p is a line l through p for which $I(p,f,l) \geq m+1$. We say that two curves f, g are *tangent at a point* p of intersection when there exists a line l which is both a tangent to f at p, and a tangent to g at p: two curves f, g are *tangent* when there exists an intersection point p at which they are tangent. Our first proposition reduces the problem of determining tangents to that of factorizing a binary form.

Lemma 7.1 *Let $p = (a,b)$ be a point of multiplicity m on an algebraic curve $f(x,y)$ in $\mathbb{K}^2$, and let $g(X,Y) = f(X+a, Y+b)$. The tangents to f at p are the linear components of the LOT in $g(X,Y)$, translated to p. In particular, when $p = (0,0)$ the tangents at p are the linear components of the LOT in $f(X,Y)$ itself.*

Proof Write $g = G_m + G_{m+1} + \cdots + G_d$ where each G_k is a binary form of degree k, and $G_m \neq 0$. Let l be the line through $p = (a, b)$ in the direction (X, Y); l is given parametrically as $x(t) = a + tX$, $y(t) = b + tY$, and is defined by $-Y(x - a) + X(y - b) = 0$. The intersection polynomial associated to the parametrization is

$$\begin{aligned}\phi(t) &= f(a + tX, b + tY) \\ &= g(tX, tY) \\ &= G_m(tX, tY) + G_{m+1}(tX, tY) + \cdots \\ &= t^m G_m(X, Y) + t^{m+1} G_{m+1}(X, Y) + \cdots.\end{aligned}$$

l is a tangent if and only if $I(p, f, l) \geq m + 1$, i.e. t^{m+1} is a factor of $\phi(t)$, i.e. $G_m(X, Y) = 0$, i.e. $-Yx + Xy$ is a linear factor of G_m. The result follows, as the translate to p is l. □

7.2 Tangents at Simple Points

Consider first the case $m = 1$ of a simple point p on a curve f. The Maclaurin expansion of the polynomial $g(X, Y)$ at the origin is

$$g(X, Y) = f_x(p)X + f_y(p)Y + \text{ HOT}$$

and there is a unique tangent line which is given by the formula

$$(x - a)f_x(p) + (y - b)f_y(p) = 0.$$

Example 7.1 Let $p = (a, b)$ be a point on the line $f = \alpha x + \beta y + \gamma$ in $\mathbb{K}^2$, so $\alpha a + \beta b + \gamma = 0$. Then $f_x = \alpha$, $f_y = \beta$ and the tangent at p to the line is defined by $(x - a)f_x(p) + (y - b)f_y(p)$, i.e. $\alpha x + \beta y + \gamma$. Thus the tangent to a line at any point is the line itself.

Example 7.2 Consider the circle $f = x^2 + y^2 - 1$ in $\mathbb{R}^2$. Singular points are given by $0 = f = x^2 + y^2 - 1$, $0 = f_x = 2x$, $0 = f_y = 2y$. Clearly, the curve is non-singular. The tangent at (a, b) on the circle has the equation $(x - a).2a + (y - b).2b = 0$, i.e. $ax + by = a^2 + b^2 = 1$. In particular, the tangent at the point $(1, 0)$ has equation $x = 1$.

Example 7.3 Here is a neat illustration of the power of elementary geometric ideas in the real affine plane. *An ellipse can be inscribed in any given parallelogram.* By this we mean that given a parallelogram, there exists an ellipse which is tangent to each of the four lines comprising the sides of the parallelogram. Suppose the parallelogram has vertices A,

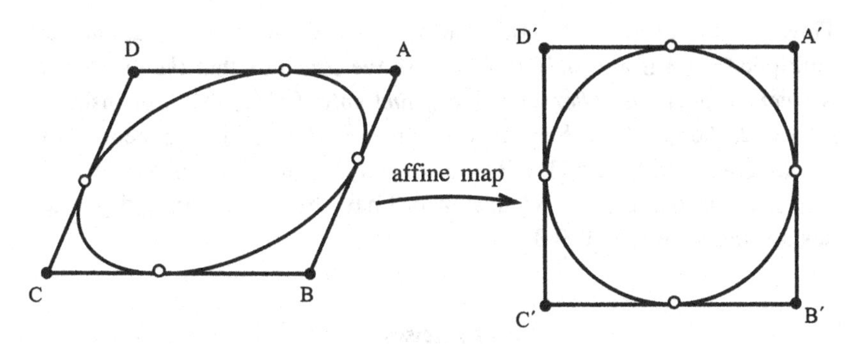

Fig. 7.1. Inscribing an ellipse in a parallelogram

B, C, D as in Figure 7.1. By the Three Point Lemma there is a unique affine map ϕ taking A, B, C respectively to $A' = (1,1)$, $B' = (1,-1)$, $C' = (-1,-1)$ and necessarily mapping D to $D' = (-1,1)$. For the parallelogram A', B', C', D' the italicized assertion is clearly true: the unit circle $x^2 + y^2 = 1$ is an inscribed ellipse, tangent to each of the four sides. Under the inverse mapping ϕ^{-1}, the circle is mapped to an ellipse. Moreover, since affine maps preserve tangency, that ellipse is likewise tangent to each of the four sides of the original parallelogram.

Let p be a simple point on a curve f in $\mathbb{K}^2$. Then the intersection number of the unique tangent line to f at p with f is ≥ 2. In general, the intersection number will be exactly two, but at exceptional points on the curve one might expect that number to take higher values. The point p is a *flex* of f when the intersection number is ≥ 3; p is an *ordinary* flex when the intersection number is $= 3$, and an *undulation* when it is ≥ 4. In Chapter 13 we will take up the question of how to find flexes, at least in principle. For the moment, we content ourselves with the very special case of curves which are graphs of polynomials, recovering a standard calculus test for their inflexional behaviour.

Example 7.4 Given a polynomial $p(x)$ over $\mathbb{K}$ the graph in $\mathbb{K}^2$ is the zero set of the curve $f(x,y) = y - p(x)$. The curve f is non-singular, and by Section 7.2 the tangent to f at the point $p = (a, p(a))$ has equation $y = (x-a)p'(a) + p(a)$. The intersections of the tangent with f at p are given by $\phi(x) = 0$ where, using the Taylor expansion of $p(x)$,

$$\phi(x) = p(x) - (x-a)p'(a) - p(a) = \frac{1}{2}(x-a)^2 p''(a) + \text{ HOT }.$$

Thus $x = a$ is a root of the equation $\phi(x) = 0$ of multiplicity ≥ 2, and has multiplicity ≥ 3 if and only if $p''(a) = 0$. We conclude that *the graph of a polynomial* $p(x)$ *has a flex at* $x = a$ *if and only if* $p''(a) = 0$. For instance for $n \geq 2$, the graph of the polynomial $p(x) = x^n$ in $\mathbb{K}^2$ has tangent $y = 0$ at $(0,0)$, and $(0,0)$ is a flex if and only if $0 = p'(0)$, i.e. if and only if $n \geq 3$. In particular, $y = x^3$ has an ordinary flex at $x = 0$, and $y = x^4$ has an undulation at $x = 0$.

Exercises

7.2.1 Let m, n be positive integers. Show that the curve $f(x,y) = x^m y^n - 1$ is non-singular. Let p be a point on f, and let q, r be the points where the tangent line at p meets the x-axis, y-axis respectively. Show that pq/pr is constant, and find its value. (pq is the distance from p to q, and pr is the distance from p to r.)

7.2.2 Find the four lines through $(0,0)$ which are tangent to the cubic curve $y^2 = x(x^2 + 7x + 1)$ in $\mathbb{R}^2$ at some other point. Sketch the curve, and the four lines.

7.2.3 Let a, b be positive real numbers. Show that the hyperbola $x^2/a^2 - y^2/b^2 = 1$ in $\mathbb{R}^2$ is non-singular, and find the equation of the tangent line at the point (α, β). Let (α, β) be a point on the hyperbola with $\beta \neq 0$. Find functions $x(\alpha)$ and $y(\beta)$ such that $(x(\alpha), 0)$ and $(0, y(\beta))$ are the respective points where the tangent line meets the x-axis and the y-axis. Find a quartic curve $f(x,y)$ in $\mathbb{R}^2$ with the property that the point $(x(\alpha), y(\beta))$ lies on the curve, for any choice of (α, β). Show that any point (x,y) on $f(x,y)$ satisfies the relation $-a < x < a$. Let (x,y) be any point on the curve $f(x,y)$ with $(x,y) \neq (0,0)$. Show that $x = x(\alpha)$ for some α with $\alpha < -a$ or $\alpha > a$. Deduce that there is a point (α, β) on the hyperbola with $\beta \neq 0$ for which $x = x(\alpha)$, $y = y(\beta)$. Sketch the zero set of the curve $f(x,y)$. (The curve f is the bullet nose of Exercise 6.3.5.)

7.3 Tangents at Double Points

Let us return to Lemma 7.1, and look more carefully at tangents to curves at singular points. We have a point $p = (a,b)$ of multiplicity m on a curve f in $\mathbb{K}^2$, and obtain the tangents as the linear factors of the LOT in the polynomial $g(X,Y) = f(X+a, Y+b)$. In the case $m = 1$ of a simple point, the LOT are linear, and we have a *unique* tangent, independently

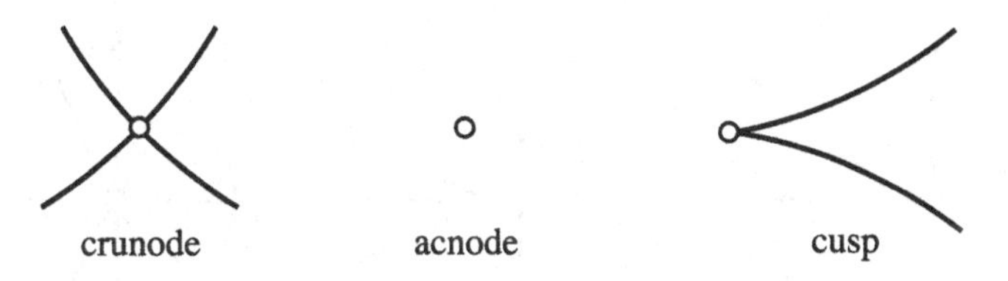

Fig. 7.2. Types of real double point

of the ground field $\mathbb{K}$. But when p is a singular point the LOT yield a form of higher degree, and the situation is more complicated. It is worth looking at the case $m = 2$ of double points in detail. In that case, the LOT have a Taylor expansion

$$g(X, Y) = X^2 f_{xx}(p) + 2XY f_{xy}(p) + Y^2 f_{yy}(p) + \text{ HOT}$$

and the tangents are obtained by factorizing the quadratic part. There are two possibilities: the tangents are distinct (and p is said to be a *node*) or the tangents coincide (and p is said to be a *cusp*); in the case of a cusp the repeated tangent is called the *cuspidal tangent*. The discriminant Δ of the quadratic part given below distinguishes nodes ($\Delta \neq 0$) from cusps ($\Delta = 0$)

$$\Delta = f_{xy}(p)^2 - f_{xx}(p) f_{yy}(p).$$

Example 7.5 The reader will readily verify that $p = (-2, 2)$ is a singular point of the sextic $f(x, y) = y^3(4 - y)^3 - 4x^4(x + 3)^2$ in $\mathbb{C}^2$. Here, writing $x = X - 2$, $y = Y + 2$, we get

$$g(X, Y) = f(X - 2, Y + 2) = -24(X^2 + 2Y^2) + \text{HOT}.$$

Thus p is a double point. The LOT in $g(X, Y)$ are $X^2 + 2Y^2$. Over the real field the LOT do not factorize, and there are no tangents. But over the complex field there are factors $X \pm i\sqrt{2}Y$, yielding distinct tangents $(x + 2) \pm i\sqrt{2}(y - 2) = 0$, so p is a node.

Over the real field there are further distinctions. Any curve f in $\mathbb{R}^2$ can be considered as a curve in $\mathbb{C}^2$, and any point p in $\mathbb{R}^2$ on f represents a real point in $\mathbb{C}^2$. Suppose p is a double point on the complex curve. Then either p is a node, or it is a cusp. In the case of a cusp, the (unique) tangent line is *real*. But in the case of a node, the tangent lines are either both real, or complex conjugate. In the former case, we say that p is a *crunode* on the real curve f, and in the latter case, that p is an *acnode*. Figure 7.2 illustrates *instances* of the three types of real double

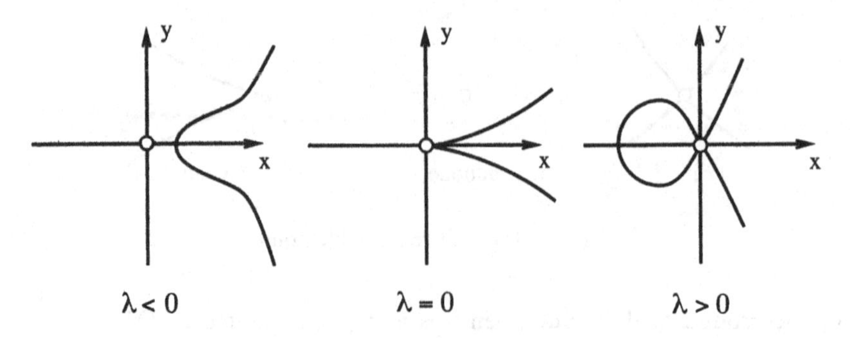

Fig. 7.3. Double points in a family of cubics

point. There is more to these pictures than meets the eye: for the acnode and crunode, it can be shown that they are 'correct', in a strictly defined technical sense, and the same is true for the cusp, provided it is 'ordinary' in the sense of Exercise 7.4.3.

Example 7.6 The cubic $y^2 = x^3 + \lambda x^2$ in $\mathbb{R}^2$ has a unique singular point $p = (0,0)$ of multiplicity 2. For $\lambda > 0$, there are distinct real tangents, and p is a crunode; for $\lambda < 0$, there are distinct complex conjugate tangents, and p is an acnode; for $\lambda = 0$, the point p is a cusp. (Figure 7.3.)

Example 7.7 The quartic $y^2 + x^4 + y^4 = 0$ in $\mathbb{R}^2$ has a unique singular point $p = (0,0)$, namely a cusp with repeated tangent $y = 0$. That may seem rather strange at first, since p is the only point in the zero set! It is better to think of f as a curve in $\mathbb{C}^2$ (so having an infinite zero set) whose tangent at p happens to have a real equation.

Exercises

7.3.1 Show that the cubic curve $f = (x+y)^3 - (x-y)^2$ in $\mathbb{R}^2$ is irreducible. Find the singular points of f, and the tangents at these points.

7.3.2 Show that the cubic curve $f = x^3 + y^3 - 3x^2 - 3y^2 + 3xy + 1$ in $\mathbb{R}^2$ is irreducible. Show that f has exactly one singular point, and find the tangents at that point.

7.3.3 Show that the cubic curve $f = y^3 - y^2 + x^3 - x^2 + 3xy^2 + 3x^2y + 2xy$ in $\mathbb{R}^2$ has exactly one singular point, and determine the tangents at that point.

7.3.4 A cubic curve f in $\mathbb{C}^2$ is given by $f = x^2y + xy(y-1) + (y-1)^2$. Show that f has exactly one singular point, at $p = (0,1)$. Show that p is a node of f, and that the tangents to f at p are $y = \omega x + 1$, $y = \omega^2 x + 1$ where ω is a primitive complex cube root of unity.

7.3.5 The *Bernouilli lemniscate* is the quartic curve in $\mathbb{R}^2$ defined by $f = (x^2+y^2)^2 - x^2 + y^2$. Show that the curve has exactly one singular point, namely a crunode at $(0,0)$. Are there any further singular points in $\mathbb{C}^2$?

7.3.6 Show that the quartic curve $f = (x^2-1)^2 - y^2(3+2y)$ in $\mathbb{R}^2$ has precisely three singular points, all of which are crunodes.

7.3.7 Show that the quartic curve $f = x^2(3y - 2x^2) - y^2(1-y)^2$ in $\mathbb{R}^2$ has two singular points, one a cusp and the other a crunode. Show that there are no further singular points.

7.3.8 In Exercise 6.3.6, it was shown that the astroid defined by the formula $f = (1 - x^2 - y^2)^3 - 27x^2y^2$ has exactly four singular points in $\mathbb{R}^2$ and exactly four further singular points in $\mathbb{C}^2$. Show that all four singular points in $\mathbb{R}^2$ are cusps, and find the tangents at these cusps. Show that the four singular points in $\mathbb{C}^2$ are nodes. (You do not need to find the tangents at the nodes.)

7.3.9 Let $\phi(x)$ be a polynomial of degree $d \geq 3$ with coefficients in $\mathbb{K}$. In Exercise 6.3.9, it was shown that any singular point of the curve $f = y^2 - \phi(x)$ in $\mathbb{K}^2$ lies on the line $y = 0$, and that the point $p = (a,0)$ is singular on f if and only if a is a repeated root of $\phi(x)$. Assuming p is singular, show that it is a double point, and is a cusp if and only if a is a zero of $\phi(x)$ of multiplicity ≥ 3.

7.4 Tangents at Points of Higher Multiplicity

Let us return once more to Lemma 7.1, and consider the general situation. We have a point $p = (a,b)$ of multiplicity m on a curve f in $\mathbb{K}^2$, and obtain the tangents as the linear factors of the LOT in the polynomial $g(X,Y) = f(X+a, Y+b)$.

Consider first the situation over the complex field, where the LOT have a unique factorization

$$G_m(X,Y) = (a_1X + b_1Y)^{\mu_1} \dots (a_sX + b_sY)^{\mu_s} \qquad (\dagger)$$

with distinct ratios $(a_k : b_k)$, and $\mu_1 + \cdots + \mu_s = m$. Then $G_m(X,Y) = 0$ if and only if $a_kX + b_kY = 0$ for some k. Thus for each linear factor

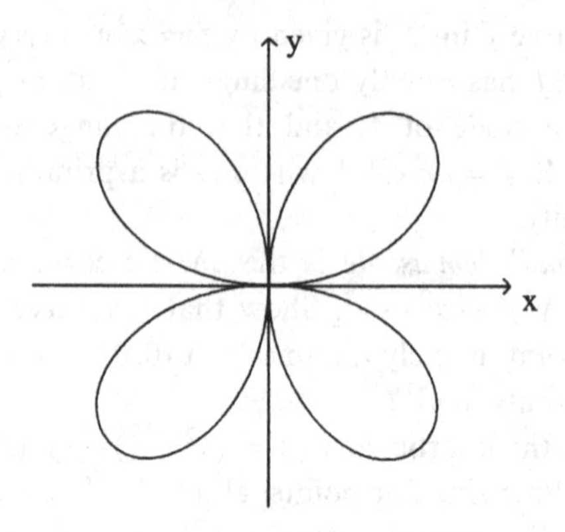

Fig. 7.4. The four-leaved clover

a_kX+b_kY of $G_m(X,Y)$, we have a unique tangent to f at p of *multiplicity* μ_k, namely the line $a_k(x-a)+b_k(y-b)=0$. Counting tangents properly (in the sense that we add up their multiplicities) we see that *over the complex field there are exactly m tangents at p, so in particular at least one*. The most general situation (in some sense) is when p is an *ordinary point* on f, i.e. it has exactly m distinct tangents. (Thus a node is an ordinary double point.)

Next consider the situation over the real field. The first remark to make is that there may be no tangents at p. (Example 7.6.) However, $G_m(X,Y)$ has a unique factorization (†) over the complex field, where the ratios $(a_k : b_k)$ are either real, or occur in complex conjugate pairs; then for each *real* ratio $(a_k : b_k)$ we have a corresponding tangent $a_k(x-a)+b_k(y-b) = 0$, and can define its 'multiplicity' as in the complex case.

Example 7.8 The sextic curve $f = (x^2+y^2)^3 - 4x^2y^2$ in $\mathbb{R}^2$ is the *four-leaved clover*. (Figure 7.4.) The curve has exactly one singular point, namely the quadruple point $p = (0,0)$ with two tangents $x = 0$, $y = 0$ both of multiplicity 2.

Example 7.9 The idea behind the concept of a 'flex' can be extended from simple to singular points. Let p be a point of multiplicity m on a curve f in $\mathbb{K}^2$. Then for any line l through p, we have $I(p,f,l) \geq m$, and l is a tangent to f at p if and only if $I(p,f,l) \geq m+1$. A tangent l at p is *ordinary* when $I(p,f,l) = m+1$, and *inflexional* when $I(p,f,l) \geq m+2$. When $m = 1$, this says that p is a 'flex'. For instance, both tangents $x = 0$,

$y = 0$ at $p = (0,0)$ to the four-leaved clover of Example 7.8 meet the curve at p with intersection number 6, so are inflexional.

Suppose we have a point p of intersection of two curves f, g in $\mathbb{K}^2$. The question arises: what are the tangents at p to $h = fg$? And the answer is given by the following proposition.

Lemma 7.2 *Let p be a point of intersection of two curves f, g in $\mathbb{K}^2$, and let $h = fg$. Then for any line l through p, the respective intersection numbers i, j, k of l at p with f, g, h are related by $k = i + j$. Moreover the tangents, to h at p comprise the tangents to f at p, and the tangents to g at p.*

Proof Since multiplicities are invariant under affine mappings, it is no restriction to suppose that $p = (0,0)$. Let l be any line through p, parametrized as $x = x(t)$, $y = y(t)$, and let $\alpha(t)$, $\beta(t)$, $\gamma(t)$ be the intersection polynomials associated respectively to f, g, h. Then

$$\begin{aligned} \gamma(t) &= h(x(t), y(t)) = fg(x(t), y(t)) \\ &= f(x(t), y(t))g(x(t), y(t)) = \alpha(t)\beta(t). \end{aligned}$$

We can write $\alpha(t) = t^i a(t)$, $\beta(t) = t^j b(t)$ where $a(0) \neq 0$, $b(0) \neq 0$. Then $\gamma(t) = t^{i+j}c(t)$ where $c(t) = a(t)b(t)$ has the property that $c(0) \neq 0$; it follows that $k = i + j$. Now let $m \leq i$, $n \leq j$ be the multiplicities of p on f, g. Then p has multiplicity $m+n$ on h, by Lemma 6.7. According to the definitions, l is a tangent at p to h if and only if $m + n + 1 \leq k = i + j$, i.e. if and only if $m + 1 \leq i$ or $n + 1 \leq j$, i.e. if and only if l is a tangent at p to f, or l is a tangent at p to g. The result follows. □

Exercises

7.4.1 Let a be a non-zero complex number. Show that the affine curve $f = y^3(4a - y)^3 - 4x^4(x + 3a)^2$ in $\mathbb{C}^2$ has five singular points, namely two triple points, a node, and two cusps.

7.4.2 Verify the statement made in Example 7.8, namely that in $\mathbb{R}^2$ the four-leaved clover $f = (x^2+y^2)^3 - 4x^2y^2$ has exactly one singular point, a quadruple point at the origin. Show that there are no further singular points in $\mathbb{C}^2$.

7.4.3 Let p be a cusp on an algebraic curve f in $\mathbb{K}^2$, with tangent line t. Show that $I(p, f, t) \geq 3$. p is said to be an *ordinary* cusp when t is an ordinary tangent. In the case when $p = (0,0)$ and t is the line $y = 0$, show that p is an ordinary cusp if and only if $f_{xxx}(0,0) \neq 0$. (Write down explicitly the quadratic and cubic terms in f.)

7.4.4 Let p be a cusp on an algebraic curve f in $\mathbb{K}^2$. Show that if p is ordinary, it is not possible to have more than one irreducible component of f passing through p. Give an example of a curve f which has a cusp p through which pass two irreducible components of f.

8

Rational Affine Curves

Lines have the rather special property that they can be parametrized by polynomial functions $x(t)$, $y(t)$, indeed by functions $x(t) = x_0 + x_1 t$, $y(t) = y_0 + y_1 t$. Likewise, the parabola $y^2 = x$ can be parametrized as $x(t) = t^2$, $y(t) = t$. However, this pattern soon breaks down.

Example 8.1 We claim that the conic $x^2 + y^2 = 1$ in $\mathbb{R}^2$ *cannot* be polynomially parametrized, in the sense that there do not exist non-constant polynomials $x = x(t)$, $y = y(t)$ with $x^2+y^2 = 1$. (We suppress the variable t for concision.) Suppose that were possible. Differentiating this identity with respect to t, we would obtain a second identity $x'x+y'y = 0$, where the dash denotes differentiation with respect to t. Think of these identities as two linear equations in x, y. Thus we obtain $xz = y'$, $yz = -x'$ where $z = xy' - x'y$. Write d, e, f for the degrees of the polynomials $x(t)$, $y(t)$, $z(t)$. Assume $z(t)$ is not zero, as a polynomial in t. Then, taking degrees in these relations, we obtain $d + f = e - 1$, $e + f = d - 1$ and hence $f = -1$, which is impossible. Thus $z(t)$ is zero, implying that $x'(t)$, $y'(t)$ are zero, and hence that $x(t)$, $y(t)$ are constant, which is the required contradiction.

If we replace 'polynomial' functions by 'rational' functions, we obtain a much more useful concept, fundamental to the theory of algebraic curves. Recall that a rational function of t is one which is a quotient $u(t)/v(t)$ of two polynomials $u(t)$, $v(t)$ with non-zero denominator $v(t)$.

Example 8.2 In $\mathbb{R}^2$ consider the circle $x^2 + y^2 = a^2$, and the point $p = (-a, 0)$ where $a > 0$. The line $x = -a$ is the tangent at p. Any other line through p has the form $y = t(x + a)$ for some scalar t and meets the circle at p, and some other point, whose coordinates depend on t. To find

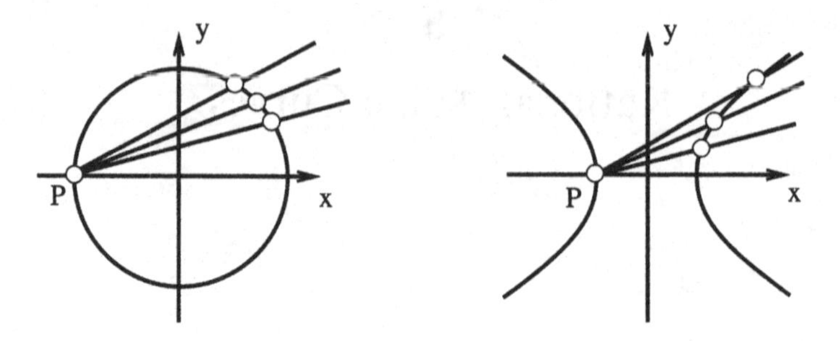

Fig. 8.1. Rationality of circle and hyperbola

the second point, substitute in the equation of the circle to get

$$(1+t^2)x^2 + 2at^2x + a^2(t^2-1) = 0.$$

$x = -a$ must be a root, so $(x+a)$ must be a factor. By inspection, the other factor is $(1+t^2)x + a(t^2-1)$, so the coordinates of the second point of intersection are

$$x(t) = \frac{a(1-t^2)}{1+t^2}, \qquad y(t) = t(x(t)+a) = \frac{2at}{1+t^2}.$$

$p = (-a, 0)$ is the *only* point on the circle that does not correspond to any value of t. Working over the complex (instead of the real) field, we obtain the same formulas. The only difference is that then $x(t)$, $y(t)$ fail to be defined when $t = \pm i$.

8.1 Rational Curves

The circle example above leads us to the following definition. Let f be an irreducible curve in $\mathbb{K}^2$. The curve f is *rational* when there exist rational functions $x(t)$, $y(t)$ satisfying the following conditions

(i) For all but finitely many values of t, the functions $x(t)$, $y(t)$ are defined and satisfy $f(x(t), y(t)) = 0$.
(ii) For any point (x, y) satisfying $f(x, y) = 0$, with a finite number of exceptions, there is a unique t for which $x = x(t)$, $y = y(t)$.

In view of our convention, that we are working either over the real or the complex field, condition (ii) implies that the zero set of a rational curve is non-empty. Here is a simple geometric construction which shows (in principle) that non-singular conics are rational. Take a fixed point $p = (\alpha, \beta)$ on the conic, and consider the 'pencil' $y = t(x-\alpha)+\beta$ of all lines through p. The construction is illustrated in Figure 8.1 for the circle and the hyperbola. Substituting the equation of the line in $f(x, y) = 0$, we get a quadratic in x (the intersection polynomial associated to that line) whose roots correspond to points where the line meets the conic. $x = \alpha$ is a root, corresponding to p, so $x - \alpha$ is a factor. The other factor can be found by inspection, and yields x in terms of t; then, substituting in $y = t(x-\alpha)+\beta$, we get y also in terms of t. There are two situations where the line fails to intersect the conic in a further point. The first is when the line is tangent to the conic at p, and the quadratic in x has $x = \alpha$ as a repeated root. The second situation is when the intersection polynomial has degree 1, i.e. the coefficient of x^2 in the quadratic vanishes; since that coefficient is itself a quadratic in t, that can only happen for at most two values of t.

Example 8.3 Consider the hyperbola $x^2 - y^2 = a^2$, and the point $p = (-a, 0)$. The line $x = -a$ is the tangent at p. Any other line through p has the form $y = t(x + a)$. Substitution yields the quadratic

$$(1 - t^2)x^2 - 2at^2x - a^2(1 + t^2) = 0.$$

$x = -a$ must be a root, so $(x + a)$ must be a factor. By inspection, the other factor is $(1 - t^2)x + a(1 + t^2)$, yielding the rational parametrization

$$x(t) = \frac{a(1 + t^2)}{1 - t^2}, \qquad y(t) = \frac{2at}{1 - t^2}.$$

$p = (-a, 0)$ is the *only* point on the hyperbola that does not correspond to any value of t. The rational functions $x(t)$, $y(t)$ are defined for all values of t *except* for $t = \pm 1$, corresponding to the lines $y = \pm(x + a)$ parallel to the 'asymptotes' $y = \pm x$. (We will have more to say about this concept in Example 12.11.)

Lemma 8.1 *Let f, g be affine equivalent curves in $\mathbb{K}^2$; then f is rational if and only if g is rational.*

Proof Since f, g are affine equivalent there exists an affine mapping $\phi(x, y)$ and a scalar $\lambda \neq 0$ for which $g(x, y) = \lambda f(\phi(x, y))$ for all x, y. Now suppose g is rational, so there exist rational functions $x(t)$, $y(t)$ satisfying conditions (i), (ii) of the above definition. Write $\phi(x, y) = (X, Y)$, where $X = px + qy + \alpha$, $Y = rx + sy + \beta$ and define $X(t)$, $Y(t)$ by $X(t) = px(t) + qy(t) + \alpha$, $Y(t) = rx(t) + sy(t) + \beta$. Then $X(t)$, $Y(t)$ are rational, and defined whenever both $x(t)$, $y(t)$ are; moreover, in that case $0 = g(x(t), y(t)) = \lambda f(\phi(x(t), y(t))) = \lambda f(X(t), Y(t))$, and hence $0 = f(X(t), Y(t))$. It remains to show that all but finitely many points (X, Y) with $f(X, Y) = 0$ have the form $X = X(t)$, $Y = Y(t)$ for some t. Any point (X, Y) with $f(X, Y) = 0$ has the form $(X, Y) = \phi(x, y)$ with $g(x, y) = \lambda f(X, Y) = 0$. And for all but finitely many (x, y), hence all but finitely many (X, Y), there exists a t with $x = x(t)$, $y = y(t)$, hence $X = X(t)$, $Y = Y(t)$. □

Example 8.4 Any irreducible complex conic is rational, since by Lemma 5.3 any such conic is affine equivalent to the parabola $y^2 = x$ or the general conic $y^2 = x^2 + 1$, both of which are rational. Likewise any irreducible real conic having a non-empty zero set is rational, since by Lemma 5.2 any such conic is affine equivalent to the parabola $y^2 = x$, the real ellipse $y^2 = -x^2 + 1$ or the hyperbola $y^2 = x^2 + 1$, all of which are rational.

The above technique for finding rational parametrizations of conics applies in principle to cubics having exactly one singular point, i.e. any line through the singular point meets the curve *at most once* elsewhere. Indeed, in principle, it applies to any irreducible curve of degree d having a singular point of multiplicity $(d - 1)$.

Example 8.5 The cubic $f = y^2 - x^3$ in $\mathbb{R}^2$ has a unique singular point at $p = (0, 0)$, namely a cusp, with tangent $y = 0$. The pencil of lines $y = tx$ through p meets the curve when the intersection polynomial $x^2(t^2 - x) = 0$, with the factor x^2 corresponding to the intersection of multiplicity 2 at p, and the remaining intersection given by $x = t^2$, $y = tx = t^3$.

Example 8.6 The *folium of Descartes*, i.e. the cubic $f = x^3 + y^3 + 3xy$ in $\mathbb{R}^2$ has a unique singular point at $p = (0, 0)$, namely a crunode with tangents $x = 0$, $y = 0$. (Figure 8.2.) The pencil of lines $y = tx$ through p meets the curve when the intersection polynomial $x^2\{(1 + t^3)x + 3t\} = 0$, with the factor x^2 corresponding to the intersection of multiplicity 2 at p, and the remaining intersection given by

$$x = \frac{-3t}{1 + t^3}, \qquad y = tx = \frac{-3t^2}{1 + t^3}.$$

Note that $t = 0$ corresponds to the tangent $y = 0$, which intersects the curve only at p, and when $t = -1$ the intersection polynomial has degree 2, and there is no further point of intersection. This corresponds to the 'asymptote' $y = 1 - x$, a concept we will discuss in Section 12.6.

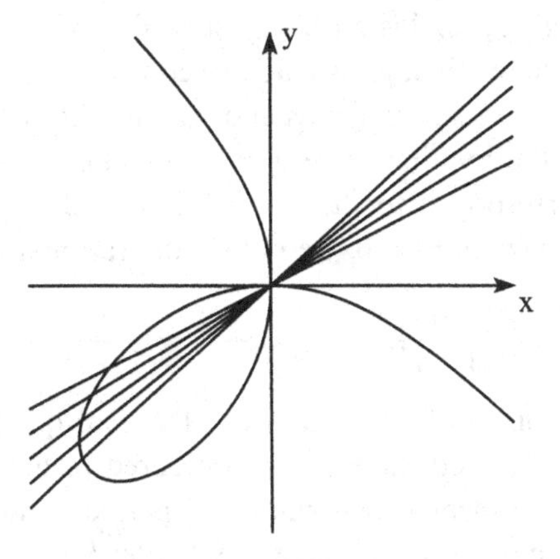

Fig. 8.2. Rationality of the folium of Descartes

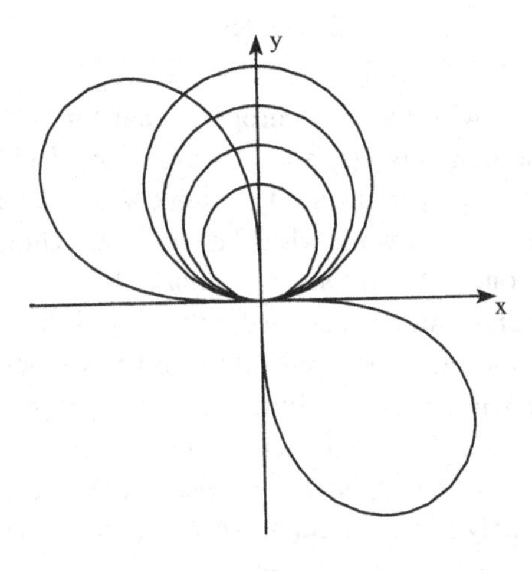

Fig. 8.3. Rationality of $xy + (x^2 + y^2)^2$

However this pattern of examples does not persist. In our final example, we present a rational quartic f in $\mathbb{R}^2$ having a unique singular point at the origin. However, rationality cannot be established via the pencil of lines through the singular point – instead we consider a 'pencil' of circles.

Example 8.7 Consider the quartic $f = xy + (x^2 + y^2)^2$ in $\mathbb{R}^2$. The curve f has a unique singular point, namely a crunode at $p = (0,0)$, the zero set is depicted in Figure 8.3. We cannot establish rationality via lines $y = tx$ through p; for instance any such line with $t < 0$ meets f in exactly *two* points other than p. Instead, we consider the 'pencil' of circles $g_t = x^2+y^2-ty$ centred on the y-axis and passing through p. Eliminating $(x^2 + y^2)$ between the relations $f = 0$, $g_t = 0$ yields $y(x + t^2y) = 0$. The solution $y = 0$ corresponds to the intersection at the origin, otherwise, substituting $y = -x/t^2$ in $g_t = 0$, we obtain the rational parametrization

$$x = \frac{-t^3}{1+t^4}, \qquad y = -\frac{x}{t^2} = \frac{t}{1+t^4}.$$

This example is on the level of a trick. The real question is why the 'pencil' of circles g_t should have been considered in the first place. The answer lies in a set of ideas which will make perfect sense once we move from the affine to the projective plane in Chapter 9.

Exercises

8.1.1 Find a rational parametrization $x = x(t)$, $y = y(t)$ of the circle $x^2 + y^2 = 5$ with the extra property that for all rational parameters t the numbers $x(t)$ and $y(t)$ are *rational*. (The point of the question is that you do need to think about which point on the circle you want to work with. The 'obvious' choices don't work; you need one with *rational* coordinates.)

8.1.2 A cubic curve f in $\mathbb{R}^2$ is defined by $f = ay^2-xy^2-ax^2-x^3$, where $a > 0$. Show that f is irreducible, and has exactly one singular point, namely a crunode. Find a rational parametrization for f.

8.1.3 A cubic curve f in $\mathbb{C}^2$ is defined by $f = ay^2 - xy^2 - x^3$ where $a > 0$. Show that f is irreducible, and has a unique singular point, namely a cusp. Find a rational parametrization for f.

8.1.4 A cubic curve f in $\mathbb{R}^2$ is given by $f = x^2y + xy(y-1) + (y-1)^2$. Show that f has exactly one singular point, namely an acnode. Construct a rational parametrization for f.

8.1.5 In Example 8.6, it was shown that the folium of Descartes, given by $f = x^3 + y^3 + 3xy$ admits the rational parametrization

$$x = \frac{-3t}{1+t^3}, \qquad y = \frac{-3t^2}{1+t^3}.$$

Let a, b, c be the parameters of three distinct points. Show that the points are collinear if and only if $abc = -1$.

8.1.6 In Example 1.8, it was shown that the pedal curve of the standard parabola $y^2 = 4ax$ where $a > 0$, with respect to the pedal point

$p = (\alpha, 0)$ satisfies the equation of the cubic $f = x(x-\alpha)^2 + y^2(a-\alpha+x)$ in $\mathbb{R}^2$. And in Exercise 6.2.1, it was shown that p is a double point of f. Find a rational parametrization for f.

8.1.7 For any real number $\lambda \neq 1$ an irreducible cubic curve f in $\mathbb{R}^2$ is defined by the formula $f(x,y) = x(x-\lambda)^2 + (x+1-\lambda)y^2$. Show that f has exactly one singular point p, whose coordinates depend on λ. Construct a rational parametrization for f.

8.1.8 Show that the quartic curve $f = x^4 - x^2y + y^3$ in $\mathbb{R}^2$ has a unique singular point, namely a triple point at the origin. Find a rational parametrization for f.

8.1.9 Show that the quartic curve $f = y(x^2+y^2) - x^4 - y^4$ in $\mathbb{R}^2$ has a unique singular point, namely a triple point at the origin. (The curve f is the boomerang illustrated in Figure 6.3.) Find a rational parametrization for f.

8.1.10 Show that the quartic curve $f = (x^2+y^2)^2 - y(3x^2-y^2)$ in $\mathbb{R}^2$ is irreducible, and has a unique singular point, namely a triple point at $p = (0,0)$. Find a rational parametrization of f.

8.1.11 Let f be the quartic curve in the real affine plane $\mathbb{R}^2$ defined by $f(x,y) = 5x^2y^2 + x^2y + xy^2 - 3x^2 + y^2$. Show that there is a conic $g(x,y)$ in $\mathbb{R}^2$ with the property that if $(x,y) \neq (0,0)$ and $f(x,y) = 0$ then $g(1/x, 1/y) = 0$. Deduce that f is rational.

8.1.12 Find a rational parametrization for the quartic curve defined by $f = x^4 - x^2y - y^3$. (Example 6.13.)

8.1.13 Find a rational parametrization for the sextic curve defined by $f = x^6 - x^2y^3 - y^5$.

8.1.14 Let f be an irreducible curve in $\mathbb{C}^2$ of degree d, having a singular point of multiplicity $(d-1)$ at some point p. Why does f have no other singular points? Show, by considering the pencil of lines through p, that f is rational. (Most of the previous exercises can be viewed as special cases of this one.)

8.1.15 Show that the quartic $f = y^2 - x^2(x^2-1)$ in $\mathbb{R}^2$ is rational, by considering its intersections with the 'pencil' of parabolas $g_t = y - tx(x-1)$.

8.2 Diophantine Equations

Conics appear in the problem of finding integral solutions of polynomial equations in several variables. The subject is named after Diophantus of Alexandria, working around the middle of the third century.

Example 8.8 In Example 2.2 we raised a classical problem of Greek mathematics, namely that of finding all right-angled triangles with integer sides, i.e. all positive integers X, Y, Z with $X^2+Y^2 = Z^2$. In this extended example, we will present a complete solution of this problem, based on the fact that the circle admits a rational parametrization.

We can assume our triangles are *primitive*, in the sense that X, Y, Z have no common factor. In that case, no two of X, Y, Z can have a common factor. If, for instance, X, Y had a common factor, they would have a common *prime* factor p which would divide Z^2, and hence Z, contradicting the assumption that the triangle is primitive. In particular, X, Y cannot both be even. What is not quite so obvious is that X, Y cannot both be odd. The proof rests on the fact that the square of an even number is congruent to 0 mod 4, whilst the square of an odd number is congruent to 1 mod 4. If X, Y were both odd then $X^2 + Y^2 \equiv 1+1 = 2$ mod 4, so $Z^2 \equiv 2$ mod 4, a contradiction. It is no restriction to suppose that Y is even.

The point (x, y) on the circle $x^2+y^2 = 1$ defined by $x = X/Z$, $y = Y/Z$ is then a *rational* point, i.e. x, y are both rational, with the fractions in their lowest terms. By Example 8.2 the circle is parametrized by

$$x = \frac{1-t^2}{1+t^2}, \qquad y = \frac{2t}{1+t^2}.$$

Clearly, if t is rational then x, y are rational. Conversely, suppose x, y are rational, then the first formula shows that t^2 is rational, and then the second formula shows that t is rational. We can therefore assume t is rational, say $t = u/v$ with u, v coprime integers, and for (x, y) to lie in the positive quadrant, we need $0 \leq t \leq 1$, i.e. $0 \leq u \leq v$. Thus all positive rational solutions are given by

$$x = \frac{X}{Z} = \frac{v^2-u^2}{v^2+u^2}, \qquad y = \frac{Y}{Z} = \frac{2uv}{v^2+u^2}$$

with u, v coprime integers satisfying $0 \leq u \leq v$. Thus there exists a positive integer λ for which

$$\lambda X = v^2 - u^2, \qquad \lambda Y = 2uv, \qquad \lambda Z = v^2 + u^2.$$

We claim that $\lambda = 1$. The integer λ divides $v^2 + u^2$, $v^2 - u^2$ and hence their sum $2v^2$, and difference $2u^2$. But u, v are coprime, so λ divides 2, and either $\lambda = 1$ or $\lambda = 2$. We will exclude the latter case. Suppose $\lambda = 2$. Since we are assuming Y is even, the middle relation shows that 4 divides $2uv$, so 2 divides uv and one of u, v is even. Now u, v cannot both be even, since they are coprime, so one is even and one is odd; it follows that one of u^2, v^2 is even and the other is odd, so $v^2 + u^2$ is odd. That contradicts the fact that 2 divides $\lambda Z = v^2 + u^2$, establishing that $\lambda = 1$, and yielding the complete solution

$$X = v^2 - u^2, \qquad Y = 2uv, \qquad Z = v^2 + u^2.$$

Here are some of the simplest primitive triangles, obtained by taking $u < v \leq 5$, the first entry providing the familiar 3,4,5 triangle of school geometry.

u	v	X	Y	Z
1	2	3	4	5
2	3	5	12	13
1	4	15	8	17
3	4	7	24	25
2	5	21	20	29
4	5	9	40	41

It is worth pondering this example a little longer. It shows that there are infinitely many rational points on the conic x^2+y^2-1. More generally, the argument shows that if we start with a conic having rational coefficients and apply the above geometric construction starting from a rational point, then we will establish the existence of infinitely many rational points on that conic. Thus (in principle) such a conic has either no rational points, or infinitely many; the former case can arise.

Example 8.9 Consider the conic $x^2 + y^2 = 3$. Suppose this has a rational point $x = X/Z$, $y = Y/Z$. (Clearly, we can suppose that the denominators are the same.) Thus $X^2 + Y^2 = 3Z^2$. We can assume that X, Y, Z have no common factor. We claim that 3 divides neither X nor Y. For if 3 divides X then, since 3 divides $3Z^2$ we see that 3 divides Y. But then 9 divides $X^2 + Y^2 = 3Z^2$, so 3 divides Z^2, and hence 3 divides Z, contrary to assumption. It follows that $X \equiv \pm 1 \bmod 3$, $Y \equiv \pm 1 \bmod 3$ so $X^2 \equiv 1 \bmod 3$, $Y^2 \equiv 1 \bmod 3$ and $X^2 + Y^2 \equiv 1 + 1 = 2 \bmod 3$, so $X^2 + Y^2$ cannot be divisible by 3, a contradiction. Thus our conic has no rational points, otherwise expressed, the zero set in $\mathbb{Q}^2$ is empty. On an intuitive level this example is very surprising, the circle manages to weave its way through all the points with rational coordinates!

Example 8.10 A famous problem of number theory is to find all positive integers X, Y, Z with $X^n + Y^n = Z^n$, where $n \geq 3$. (The Fermat

Problem.) Equivalently, one seeks positive rational points $x = X/Z$, $y = Y/Z$ on the *Fermat* curve $x^n + y^n = 1$, of degree n. The curve is named after the French mathematician Fermat (1601-1665) who wrote in the margin of one of his books that he had found a 'truly marvellous proof' that for $n \geq 3$ the equation has no positive integer solutions. For over three centuries mathematicians tried to discover Fermat's 'proof', without success. The technique of Example 8.8 no longer works, since (as will be shown in the final chapter) the curve cannot be rationally parametrized for $n \geq 3$. As the years passed, larger and larger integers N were found with the property that Fermat's Conjecture held for all $n < N$; indeed, in recent decades, a small army of computer scientists dedicated themselves to Fermat's 'conjecture'. In 1983, the German mathematician Faltings established general results, one of whose consequences is that for a given $n \geq 4$, there are at most finitely many (positive) rational points on the Fermat curve. And finally in 1996, the English mathematician Wiles showed that Fermat's conjecture does indeed hold.

Example 8.11 We can pursue the idea of rational parametrization to find rational solutions of some Diophantine equations of higher degree. Consider, for instance, the cubic $y^2 = x^3 + x^2$, having just one singular point, namely a crunode at the origin. Parametrizing this via the pencil of lines $y = tx$ through the singular point we obtain the rational parametrization $x = t^2 - 1$, $y = t^3 - t$. Clearly, by taking rational values of t we obtain points on the cubic with rational coordinates, and conversely all such points arise in this way. But more is true. Taking integer values of t we obtain points (x, y) on the cubic with integer coordinates. Conversely, let $(x, y) \neq (0, 0)$ be a point on the cubic with integer coordinates. Then the line $y = tx$ joining it to the origin has a rational slope t. But then $x = t^2 - 1$, so t^2 is an integer, and hence t is an integer as well.

This approach to finding rational points depends on the fact that the curve can be rationally parametrized. The cubic of the next example *cannot* be rationally parametrized, though the reader will have to wait till the final chapter for a proof; however, that does not stop us finding some rational points.

Example 8.12 On the non-singular cubic $y^2 = x^3 + 17$ one can certainly spot some 'obvious' rational points, such as $a = (4, 9)$, $b = (2, 5)$, $c = (-1, 4)$. But enthusiasm for this approach is likely to wane long before you reach such solutions as $(5234, 378661)$, indeed you need an 11–digit calculator even to check this is a solution! A more profitable philosophy is that of constructing new rational points from old, via the classical *secant method* of taking the line through two given rational points and looking for a third point where the line meets the cubic. For instance, the line joining a, b is $y = 2x + 1$, and eliminating y between this and the curve yields the cubic $x^3 - 4x^2 - 4x + 16 = 0$. This is satisfied by $x = 4$, $x = 2$; the third root is determined by the fact that the sum of the roots is 4, giving $x = -2$. Substituting back in the line we get $y = -3$, giving us a new rational point $d = (-2, -3)$. It is clear from the method that it is general: if the line joining two distinct rational points on a cubic curve (with rational coefficients) produces a third point, then that point is rational. It was a fluke that d has integer coefficients, the line joining b, c produces the rational point $e = (-8/9, 109/27)$.

The secant method is attractive from the computational viewpoint, in the sense that it can be implemented on a computer. It is also attractive from the algebraic viewpoint, an idea we will pursue in Chapter 17, in the projective context. Of course for the secant method to produce anything new we need to start from at least two rational points, as in the above example. What if we can only spot one rational point? The way forward is the *tangent method* whereby the secant joining two points is replaced by the tangent line at a single point.

Example 8.13 On the non-singular cubic $y^2 = x^3 + 2x - 2$, one can spot the rational points $(1, \pm 1)$. The secant method yields nothing, since the line joining these points is $x = 1$, which meets the curve *only* at these points. However the tangent method is productive. The tangent at $a = (1, 1)$ is $5x - 2y - 3 = 0$, and eliminating y between this and the curve we obtain the cubic $4x^3 - 25x^2 + 38x - 17 = 0$. We know that $x = 1$ is a repeated root, and the third root is determined by the fact that the sum of the roots is $25/4$, yielding $x = 17/4$: substituting back in the tangent equation gives $y = 73/8$, producing a new rational point $(17/4, 73/8)$.

And what happens when one cannot spot a rational point? Surprisingly, at the time of writing no algorithm is known for deciding whether a cubic (with rational coefficients) possesses a rational point.

Exercises

8.2.1 In Exercise 8.1.13, we found a rational parametrization for the sextic curve $f = x^6 - x^2y^3 - y^5$. Use this rational parametrization to find all integer solutions of the equation $f(x, y) = 0$.

8.2.2 Spot a rational point on the cubic $y^2 = x^3 - 3x^2 + 3x + 1$, and use the tangent method to find at least one more rational point.

8.2.3 Spot a rational point on the cubic $x^3 - 3x^2y + 5y^2 - 2y = 0$, and use the tangent method to find at least one more rational point.

8.3 Conics and Integrals

One of the early aims of integration theory was to find explicit formulas for integrals $\int_a^b r(t)\,dt$ of the functions $r(t)$ which arose naturally in physical problems. It was realized that this could be achieved for the class of real rational functions $r(t) = u(t)/v(t)$ with $u(t)$, $v(t)$ real polynomials in t. Indeed it is proved in algebra texts that any rational function can be expressed as a linear combination of 'partial fractions' of the forms

$$t^p, \quad \frac{1}{(t-a)^p}, \quad \frac{bt+c}{(t^2+dt+e)^q}$$

where $t^2 + dt + e$ is irreducible. All such expressions can be explicitly integrated (the methods are presented in calculus texts) so the same is true for any rational function $r(t)$. However, numerous physical problems give rise to integrals which, although they fail to have rational integrands, are at least capable of being *reduced* to such integrals. A simple example is

$$\int_a^b \frac{dx}{\sqrt{1+x^2}}.$$

It is an elementary exercise to check that the substitution $x = 2t/(1-t^2)$ reduces the integral to one of the form

$$\int_c^d \frac{2\,dt}{1-t^2}$$

with a rational integrand. In fact (as calculus texts point out) this substitution applies to a much wider class of integrals; it will reduce any integral

$$\int_a^b R(x,y)\,dx \qquad (\star)$$

where $R(x,y)$ is a rational function of x, y, and $y = \sqrt{1+x^2}$, to one with a rational integrand. In the above example $R(x,y) = 1/y$.

Although the rational substitution $x = 2t/(1-t^2)$ is presented in calculus texts as a 'trick' for evaluating integrals, it has a very clear underlying geometry. Firstly x, y satisfy the equation of a hyperbola

$y^2 - x^2 = 1$, and secondly, the hyperbola is a rational curve. Using the pencil of lines $y = tx + 1$ through the point $(1, 0)$ we see that the 'upper' branch of the hyperbola is parametrized by the formulas below, with $-1 < t < 1$,

$$x = \frac{2t}{1-t^2}, \qquad y = \frac{1+t^2}{1-t^2}.$$

Having grasped the geometric idea we see that, in principle, we ought to be able to handle any integral (*) where $R(x, y)$ is a rational function of x, y and $y = \sqrt{\alpha x^2 + \beta x + \gamma}$. Indeed x, y are then related by $y^2 = \alpha x^2 + \beta x + \gamma$ defining a real conic which can be rationally parametrized, at least provided it is irreducible and has real points. More generally, this argument will apply to *any* such conic $R(x, y)$, for by Lemma 5.2 it can be put into the shape $y^2 = \alpha x^2 + \beta x + \gamma$ by affine transformations.

9
Projective Algebraic Curves

One of the main functions of this book is to place algebraic curves in a natural setting (the complex projective plane) where they can be studied easily. For some readers, particularly those whose background is not in mathematics, this may prove to be a psychological barrier. I can only assure them that the reward is much greater than the mental effort involved. History has shown that placing algebraic curves in a natural setting provides a flood of illumination, enabling one much better to comprehend the features one meets in everyday applications.

Let me motivate this by looking briefly at the important question of understanding how two curves intersect each other. The simplest situation is provided by two lines. In general, you expect two distinct lines to meet in a single point. But there are lines ($x = 0$ and $x = 1$ for instance) which do not intersect. This really is a nuisance. It would be much nicer to say that these lines intersect at a single point, namely a 'point at infinity'. In this chapter, we will extend the ordinary affine plane to a 'projective plane' by adding points at infinity, and gain much as a result.

9.1 The Projective Plane

The construction is based on the following intuition. Think of the affine plane $\mathbb{K}^2$ as a plane L in $\mathbb{K}^3$ which does not pass through the origin O, and let L_0 be the parallel plane through O. Then we have a natural one-to-one correspondence between points in $\mathbb{K}^2$, and lines through the origin O in $\mathbb{K}^3$ not in L_0. To a point in L we assign the unique line through O and that point, and conversely, to a line through O not in L_0 we assign the point where it meets L. (Figure 9.1.) In this way, we conceive of lines through O in $\mathbb{K}^3$ in L_0 as in some sense 'points at infinity' in $\mathbb{K}^2$. That motivates the following formal definitions.

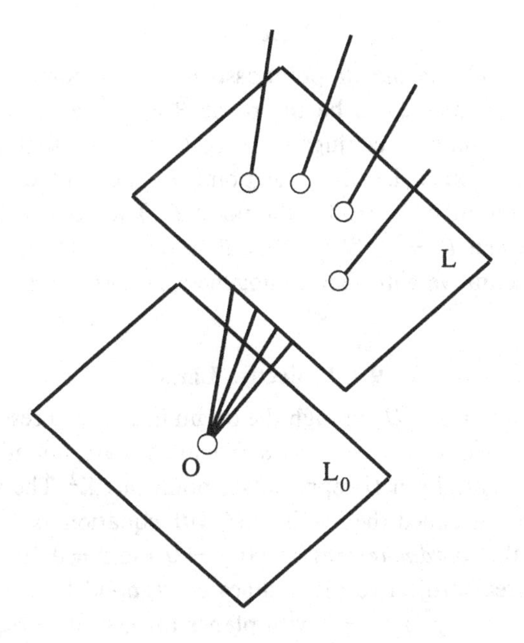

Fig. 9.1. Affine plane in 3-space

The *projective plane* $\mathbb{PK}^2$ is the set of lines through the origin O in $\mathbb{K}^3$. These lines are called the *points* of $\mathbb{PK}^2$. Given a non-zero vector $X = (x_1, x_2, x_3)$ in $\mathbb{K}^3$ there is a unique line through X, O determining a unique point in $\mathbb{PK}^2$, denoted by $(x_1 : x_2 : x_3)$. Beware, there is no point $(0 : 0 : 0)$! We call X a *representative*, and x_1, x_2, x_3 *homogeneous coordinates* for the point. Homogeneous coordinates are not unique; for any non-zero scalar λ, the vector λX represents the *same* point in $\mathbb{PK}^2$. Thus, given two points $A = (a_1 : a_2 : a_3)$, $B = (b_1 : b_2 : b_3)$ in $\mathbb{PK}^2$ the relation $A = B$ means that there exists a scalar $\lambda \neq 0$ for which $b_1 = \lambda a_1$, $b_2 = \lambda a_2$, $b_3 = \lambda a_3$. For many practical purposes we will usc homogeneous coordinates labelled x, y and z.

Example 9.1 The *sphere model* of $\mathbb{PR}^2$. Each line through the origin in $\mathbb{R}^3$ meets the unit sphere $x^2 + y^2 + z^2 = 1$ in two diametrically opposite points: so $\mathbb{PR}^2$ can be thought of as a sphere, with diametrically opposite points identified. We will expand on this model of the real projective plane as we proceed.

Exercises

9.1.1 The *complex conjugate* point associated to a point $P = (a : b : c)$ in $\mathbb{PC}^2$ is defined to be the point $\bar{P} = (\bar{a} : \bar{b} : \bar{c})$ where $\bar{a}$, $\bar{b}$, $\bar{c}$ are the complex conjugates of a, b, c. Show that this concept is well-defined, i.e. that the point $\bar{P}$ does not depend on the representative chosen for the point P. The point P is said to be *real* when $P = \bar{P}$. Show that P is real if and only if P has a representative with real homogeneous coordinates.

9.2 Projective Lines

Now consider any plane U through the origin in $\mathbb{K}^3$, so necessarily having an equation $ax + by + cz = 0$, where a, b, c are scalars, not all zero. Every line through the origin in U represents a point in $\mathbb{PK}^2$. The resulting set of points in $\mathbb{PK}^2$ is called the *line* in $\mathbb{PK}^2$ with equation $ax + by + cz = 0$. Examples are the *coordinate lines* $x = 0$, $y = 0$ and $z = 0$. In the spherical model of the real projective plane, lines correspond to intersections of the unit sphere $x^2 + y^2 + z^2 = 1$ with planes through the origin, hence to great circles.

Example 9.2 A set of points in $\mathbb{PK}^2$ is *collinear* when they all lie on one line. The condition for three points $P_1 = (x_1 : y_1 : z_1)$, $P_2 = (x_2 : y_2 : z_2)$, $P_3 = (x_3 : y_3 : z_3)$ to be collinear is easily obtained. It is the condition for the non-zero vectors P_1, P_2, P_3 in $\mathbb{K}^3$ to lie on a single plane through the origin, i.e. for those vectors to be linearly dependent, i.e. for the determinant

$$\begin{vmatrix} x_1 & y_1 & z_1 \\ x_2 & y_2 & z_2 \\ x_3 & y_3 & z_3 \end{vmatrix} = 0.$$

Example 9.3 Through any two distinct points in $\mathbb{PK}^2$ there is a unique line. That is simply the statement that through two distinct lines through the origin in $\mathbb{K}^3$ there is a unique plane through the origin. To find the equation of the line through two distinct points with representatives $P_1 = (x_1, y_1, z_1)$, $P_2 = (x_2, y_2, z_2)$ we proceed as follows. We seek all representatives $P = (x, y, z)$ for which P_1, P_2, P are coplanar; by Example 9.2 the condition is

$$\begin{vmatrix} x_1 & y_1 & z_1 \\ x_2 & y_2 & z_2 \\ x & y & z \end{vmatrix} = 0.$$

For instance, if $P_1 = (0:0:1)$, $P_2 = (1:0:1)$ then the line joining P_1, P_2 is $y = 0$.

The next example describes the crucial property which distinguishes the affine from the projective plane.

Example 9.4 Any two distinct lines in $\mathbb{PK}^2$ intersect in a unique point. Indeed, two distinct lines in $\mathbb{PK}^2$ correspond to two distinct planes through the origin in $\mathbb{K}^3$, so intersect in a unique line through the origin in $\mathbb{K}^3$, i.e. in a unique point in $\mathbb{PK}^2$. It is this property which marks the fundamental difference between the affine plane and the projective plane. A practical question is to find the point of intersection of two distinct lines $ax+by+cz = 0$ and $a'x+b'y+c'z = 0$. By linear algebra, a solution of these equations is (X, Y, Z) where

$$X = \begin{vmatrix} b & c \\ b' & c' \end{vmatrix}, \quad Y = -\begin{vmatrix} a & c \\ a' & c' \end{vmatrix}, \quad Z = \begin{vmatrix} a & b \\ a' & b' \end{vmatrix}.$$

Note that at least one of X, Y, Z is $\neq 0$, else (by linear algebra) the vectors (a, b, c), (a', b', c') are linearly dependent, and the lines coincide.

At this point, it is worth pointing out that $\mathbb{PK}^2$ is a 'projective plane' in the sense of Section 2.3. The three points $(1:0:0)$, $(0:1:0)$, $(0:0:1)$ are not collinear; Example 9.3 tells us that through any two distinct points there is a unique line, and Example 9.4 tells us that any two lines intersect in at least one point.

Example 9.5 Any projective line $ax + by + cz = 0$ in $\mathbb{PK}^2$ can be 'rationally parametrized'. (A formal treatment of these ideas will be given in Chapter 18.) Recall that a line in $\mathbb{PK}^2$ corresponds to a plane through the origin in $\mathbb{K}^3$, and that any two *distinct* points A, B on the line correspond to basis vectors for that plane. It follows that any point on the plane has the form $X = sA + tB$ for some scalars s, t. Writing $A = (a_1, a_2, a_3)$, $B = (b_1, b_2, b_3)$, $X = (x, y, z)$ that means we can write $x = sa_1 + tb_1$, $y = sa_2 + tb_2$, $z = sa_3 + tb_3$ for some ratio $(s : t)$. For instance, the line $z = 0$ passes through the points $A = (1 : 0 : 0)$, $B = (0:1:0)$ yielding the parametrization $x = s$, $y = t$, $z = 0$. As in the affine case, the parametrization is not unique, it depends on the choice of A, B.

A few words are in order on the question of 'visualizing' the zero set of a projective line. Recall first the affine case. In the real case, we have the familiar lines in the $\mathbb{R}^2$; and in the complex case, we saw in Example 2.13

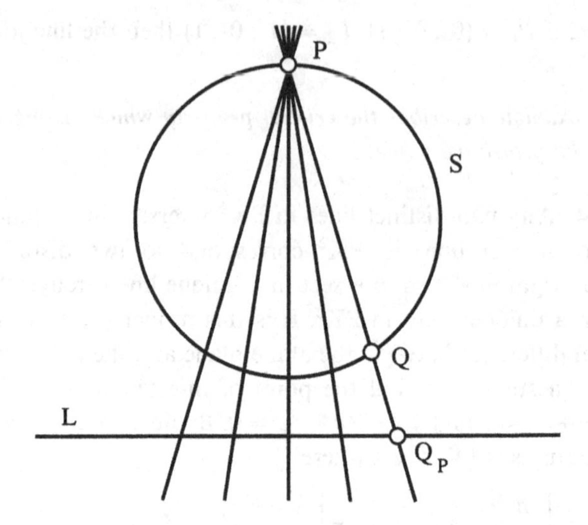

Fig. 9.2. Real projective line as a circle

that a line in $\mathbb{C}^2$ could be thought of as a real plane. Consider now the projective case, where a line is defined by a plane through the origin in $\mathbb{K}^3$. It will be no restriction to take the plane $z = 0$, so that the points of the (projective) line correspond to the (affine) lines through the origin in the (x, y)-plane, and are determined by the ratio $(x : y)$. Provided $x \neq 0$ we can identify the ratio with the quotient y/x in $\mathbb{K}$. Thus we can view the points of the projective line as the elements of $\mathbb{K}$ plus an extra element ∞. In the real and complex cases, this gives rise to simple visualizations.

Example 9.6 In the real affine plane, let S be a circle, let P be a point on S, and let L be a line not passing through P. We think of L as a copy of $\mathbb{R}$. Define *stereographic projection* to be the natural mapping $S - \{P\} \to L$, which takes a point $Q \neq P$ on S to the point on L where it meets the line through P, Q. (Figure 9.2.) Under this mapping the points of $S - \{P\}$ correspond one-to-one with the elements of $\mathbb{R}$, and we can extend the bijection to the whole of S by mapping P itself to the element ∞. Thus *we can think of a real projective line as a circle.*

Example 9.7 In real affine space, let S be a sphere, let P be a point on S, and let L be a plane not passing through P. We think of L as a copy of $\mathbb{C}$. Here too we can define *stereographic projection* to be the natural mapping $S - \{P\} \to L$, which takes a point $Q \neq P$ on S to the point on

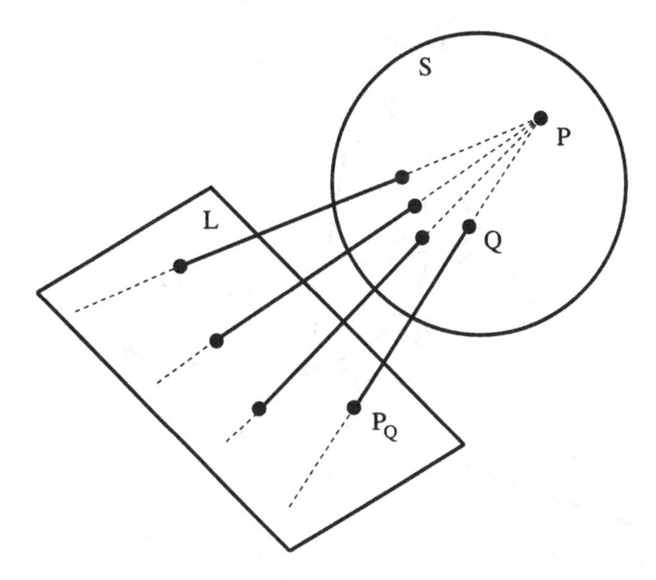

Fig. 9.3. Complex projective line as a sphere

L where it meets the line through P, Q. (Figure 9.3.) Under this mapping the points of $S - \{P\}$ correspond one-to-one with the elements of $\mathbb{C}$, and we can extend the bijection to the whole of S by mapping P itself to the element ∞. Thus *we can think of a complex projective line as a sphere.* This continues the train of thought begun in Example 2.13, namely that *'curves' in the complex projective plane can be thought of as real surfaces.*

Example 9.8 Just for the fun of it, consider the projective plane $P\mathbb{Z}_2^2$ over the field $\mathbb{Z}_2$ with two elements 0, 1. There are seven points: $(0:0:1)$, $(1:0:1)$, $(0:1:1)$, $(1:1:1)$, $(1:0:0)$, $(0:1:0)$, $(1:1:0)$. And there are seven lines: $x = 0$, $y = 0$, $z = 0$, $x = y$, $y = z$, $z = x$, $z = x + y$. Think of it as in Figure 9.4.

Exercises

9.2.1 In each of the following cases, find the equation of the line joining the points A, B in $P\mathbb{C}^2$.

(i) $A = (1:0:-1)$ $B = (1:1:-2)$
(ii) $A = (1:0:0)$ $B = (1:2:3)$
(iii) $A = (-1:1:1)$ $B = (2:3:3)$
(iv) $A = (1:i:0)$ $B = (1:-i:0)$.

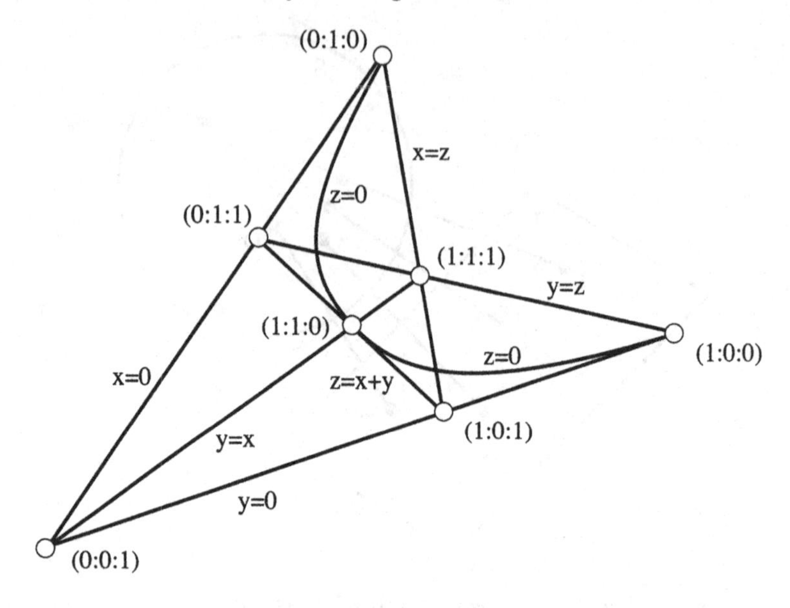

Fig. 9.4. The projective plane over $\mathbb{Z}_2$

9.2.2 In each of the following cases find the point of intersection of the two given lines in $P\mathbb{C}^2$.

(i) $2x + 3y + z = 0, \quad x - y + 2z = 0$
(ii) $x + y + z = 0, \quad x + y + 2z = 0$
(iii) $x + 3y - 2z = 0, \quad y + z = 0.$

9.2.3 The *complex conjugate* line associated to a line $L = ax + by + cz$ in $P\mathbb{C}^2$ is the line $\bar{L} = \bar{a}x + \bar{b}y + \bar{c}z$. Show that this concept is well-defined, i.e. that the line $\bar{L}$ does not depend on the equation of L. L is said to be *real* when $L = \bar{L}$. Show that L is real if and only if it has a defining equation with real coefficients.

9.2.4 Let P be a non-real point in $P\mathbb{C}^2$. Show that the line joining P, $\bar{P}$ is real. Let L be a non-real line in $P\mathbb{C}^2$. Show that the point of intersection of L, $\bar{L}$ is real.

9.2.5 Nine points P_{ij} in $P\mathbb{C}^2$ are defined as follows, where ω is a primitive complex cube root of unity. (These points will arise naturally in a geometric context in Example 13.6.)

$$\begin{array}{lll}
P_{00} = (0:-1:1) & P_{01} = (0:-\omega:1) & P_{02} = (0:-\omega^2:1) \\
P_{10} = (1:0:-1) & P_{11} = (1:0:-\omega) & P_{12} = (1:0:-\omega^2) \\
P_{20} = (-1:1:0) & P_{21} = (-\omega:1:0) & P_{22} = (-\omega^2:1:0).
\end{array}$$

Any two of these points determine a line. Show that in total there are twelve such lines, and that each one passes through exactly three of the points P_{ij}. You might find it interesting to form a diagram of these nine points and twelve lines, and then compare it with that in Figure 2.2.

9.2.6 Show that the projective plane $P\mathbb{Z}_3^2$ has precisely 13 points, and precisely 13 lines. How many points lie on a line? And how many lines pass through a point?

9.3 Affine Planes in the Projective Plane

We need to say something about the fundamental matter of how affine planes 'sit' in projective planes. To this end, we return to the basic intuition of Section 9.1. We start from an arbitrary affine plane L in $\mathbb{K}^3$, not through the origin O, and denote by L_0 the parallel plane through O. That gave us a natural one-to-one correspondence between points in L, and lines in $\mathbb{K}^3$ through O not lying in the plane L_0. In the language of the preceding section, L_0 defines a line in $P\mathbb{K}^2$, the *line at infinity* associated to L, whose points are the *points at infinity*; explicitly, if L has equation $ax + by + cz = d$ then L_0 has equation $ax + by + cz = 0$. Thus we obtain a natural one-to-one correspondence between points in L and points in $P\mathbb{K}^2$ not lying on the line L_0. In other words, we can think of the projective plane $P\mathbb{K}^2$ as the 'affine' plane L, plus an associated 'line at infinity' L_0. That is the underlying geometric picture. The reader should bear in mind that the 'line at infinity' depends on a choice of plane L, there is nothing unique about it.

For practical purposes, we need to be more explicit about exactly how we identify L with the affine plane $\mathbb{K}^2$. The idea is that, provided L is not parallel to a given coordinate axis, we can identify L with the corresponding coordinate plane (obtained by setting the coordinate equal to zero) by projection onto that plane, in the direction of the axis. The correspondence could be written down quite explicitly, by solving the equation $ax + by + cz = d$ for one of x, y, z in terms of the other two. For instance the affine plane L defined by $x + y + z = 1$ is not parallel to the z-axis, and can be identified with the coordinate plane $z = 0$ via $(x, y) \sim (x, y, 1 - x - y)$. For most practical purposes, it will be sufficient to take L to be one of the planes $x = 1$, $y = 1$, $z = 1$, with associated lines at infinity $x = 0$, $y = 0$, $z = 0$. These choices give rise to three basic identifications:

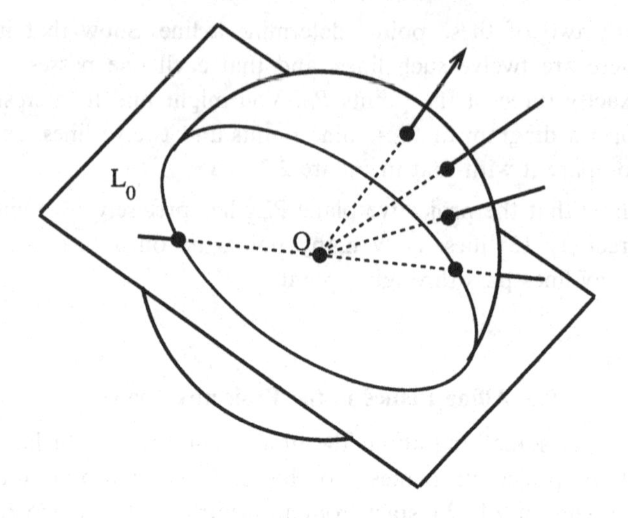

Fig. 9.5. Hemisphere model of $\mathbb{PR}^2$

choice	identification
$x = 1$	$(y, z) \sim (1 : y : z)$
$y = 1$	$(z, x) \sim (x : 1 : z)$
$z = 1$	$(x, y) \sim (x : y : 1)$

Example 9.9 The hemisphere model of $\mathbb{PR}^2$. Suppose we are given a plane L_0 through the origin O in $\mathbb{R}^3$. Each line through O meets the closed unit hemisphere $x^2 + y^2 + z^2 = 1$, $L_0 \geq 0$, with boundary a great circle, at least once. More precisely, every line through O meets the open hemisphere $x^2 + y^2 + z^2 = 1$, $L_0 > 0$ just *once*, whilst lines which meet the boundary great circle $x^2 + y^2 + z^2 = 1$, $L_0 = 0$ do so *twice*. Thus, in contrast to the sphere model, we have only to identify diametrically opposite points *on the boundary*. Perhaps this represents the best visualization of $\mathbb{PR}^2$, with the open hemisphere corresponding to $\mathbb{R}^2$, and the boundary circle to a line. Looking at the hemisphere from a distant point on the line perpendicular to L_0 through O one 'sees' the projective plane as a disc. (Figure 9.5.) Points in the interior correspond one-to-one with points in the real affine plane $\mathbb{R}^2$; the boundary circle corresponds to a line, but 'opposite' points have to be identified. It is conceptually important to realize that, whereas the sphere model provides

us with a *single* representation of the real projective plane, the hemisphere model provides a whole *family* of representations, one for each choice of the plane L_0.

9.4 Projective Curves

In this section we will work in the projective plane $P\mathbb{K}^2$, with homogeneous coordinates x, y, z. By a *projective algebraic curve* of degree d in $P\mathbb{K}^2$ we mean a non-zero form $F(x, y, z)$ of degree d, *up to multiplication by a non-zero scalar*. As in the affine case, we will abbreviate the term 'algebraic curve' to 'curve'. Projective curves of degree 1, 2, 3, ... are called *lines*, *conics*, *cubics*, and so on. Over the real field we refer to *real* projective curves, and over the complex field to *complex* projective curves. Given a curve F we can associate to it the set

$$C(F) = \{(x, y, z) \in \mathbb{K}^3 : F(x, y, z) = 0\}.$$

This set has the crucial property of being a *cone*, meaning that if (x, y, z) is a point in $C(F)$ then so too is (tx, ty, tz), for any choice of scalar t; that is immediate from the defining property $F(tx, ty, tz) = t^d F(x, y, z)$ of forms. (Lemma 3.12.) In view of this fact, it makes sense to define the *zero set* of F to be the set $V(F)$ defined below; the key fact is that whether a point belongs to $V(F)$ does not depend on the choice of representative for that point.

$$V(F) = \{(x : y : z) \in P\mathbb{K}^2 : F(x, y, z) = 0\}.$$

Recall from Section 3.4 that any form F over the field $\mathbb{K}$ admits a unique factorization $F = cF_1^{r_1} \dots F_s^{r_s}$ with $c \neq 0$ in $\mathbb{K}$, and the F_i irreducible forms such that F_j is not a factor of F_k for $j \neq k$. When F is a projective curve, we refer to $F_1, \dots, F_s$ as the *irreducible components* of F. And, as in the affine case, the zero set of F is the union of the zero sets of $F_1, \dots, F_s$.

Example 9.10 Let F be a curve in the real projective plane $P\mathbb{R}^2$. We can think of the zero set in terms of the sphere model of the projective plane. Since the set $C(F)$ is a cone, the zero set is completely determined by the 'curve' of intersection with the sphere $x^2 + y^2 + z^2 = 1$. Thus we can think of the zero set as a 'curve' on the sphere, with diametrically opposite points identified. (Figure 9.6.) For instance, when F is a line the cone is a plane through the origin, intersecting the sphere in a great circle. Likewise, we can think of the zero set in terms of the hemisphere

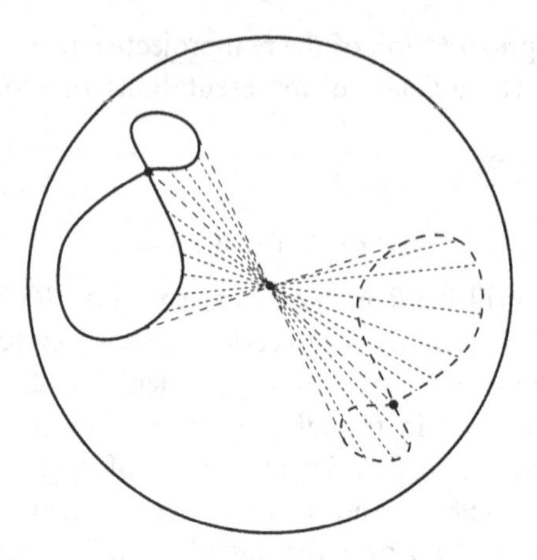

Fig. 9.6. Projective curve in the spherical model

model, where it is completely determined by the 'curve' of intersection with the hemisphere $x^2 + y^2 + z^2 = 1$, $z \geq 0$. Thus we can think of the zero set as a 'curve' on the hemisphere, with diametrically opposite points on the boundary great circle identified. For instance, when F is a line the resulting cone intersects the hemisphere in a great semicircle, and we identify the two end points.

9.5 Affine Views of Projective Curves

The object of this section is to establish the basic relations between affine and projective curves. Suppose we have a projective curve F, and make a choice of affine plane L with equation $ax + by + cz = d$. Then by eliminating one of the variables x, y, z between the relations $F(x, y, z) = 0$, $ax + by + cz = d$ we obtain an affine curve f in the (y, z)-plane, the (x, z)-plane or the (x, y)-plane. We assume here that the coefficient of the variable in the equation of L is non-zero. We refer to f as an *affine view* of F. Thus, to any projective curve F, we associate a family of affine views f, one for each choice of affine plane L and each choice of variable.

Example 9.11 Consider the real projective conic $F = x^2 + y^2 - z^2$. The affine views of F corresponding to the planes $z = 1$, $x + z = 1$ and $x + y + z = 1$ are the real ellipse $x^2 + y^2 = 1$, the parabola $y^2 = 1 - 2x$, and the hyperbola $2xy = 2x + 2y - 1$. This example makes the point

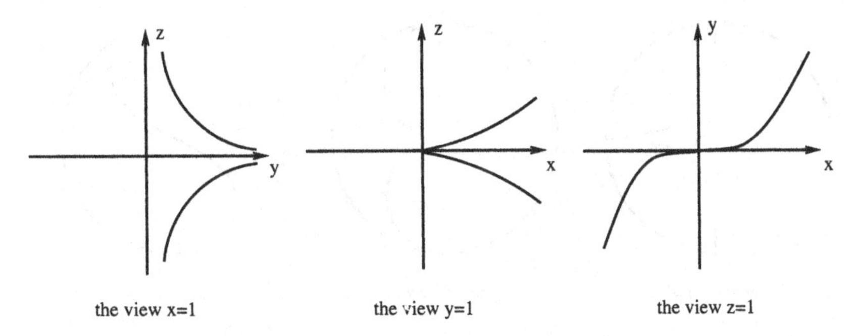

Fig. 9.7. Three affine views of a cubic curve

that different affine views of a projective curve may well fail to be affine equivalent.

Perhaps the best mental picture is to start from the spherical model of the real projective plane, with F defining a 'curve' on the surface of the unit sphere. An affine view is then determined by an affine plane L, with L_0 the parallel plane through O. The plane L_0 then determines a hemispherical model of the real projective plane, with the curve on the sphere giving a curve on the open hemisphere, and opposite points of intersection with the great circle $L_0 = 0$ identified. For instance, when F is a line, the intersection with the sphere is a great circle, intersecting the hemisphere in 'half' a great circle, with its end points identified. As L_0 changes, so the view of the 'curve' will change. In practice, we will work with the three *principal* affine views, meaning those associated to the affine planes $x = 1$, $y = 1$, $z = 1$. (Note that *any* point in the projective plane has a representative with either $x = 1$, $y = 1$ or $z = 1$ so these views cover the whole plane.) The process of obtaining the equation for f by substituting $x = 1$, $y = 1$ or $z = 1$ in $F(x, y, z)$ is known as *dehomogenization*, since it destroys the homogeneity of F.

Example 9.12 Consider, for instance, the projective cubic $F = yz^2 - x^3$ in $P\mathbb{R}^2$. Dehomogenizing with respect to x, y, z we obtain the three affine views of Figure 9.7, namely $f = y - x^3$ in the (x, y)-plane, $g = yz^2 - 1$ in the (y, z)-plane, and $h = z^2 - x^3$ in the (z, x)-plane. No two of these affine views are affine equivalent.

It is illuminating to think of Example 9.12 in terms of hemispherical models. If we take the hemispheres resulting from the planes $x = 0$,

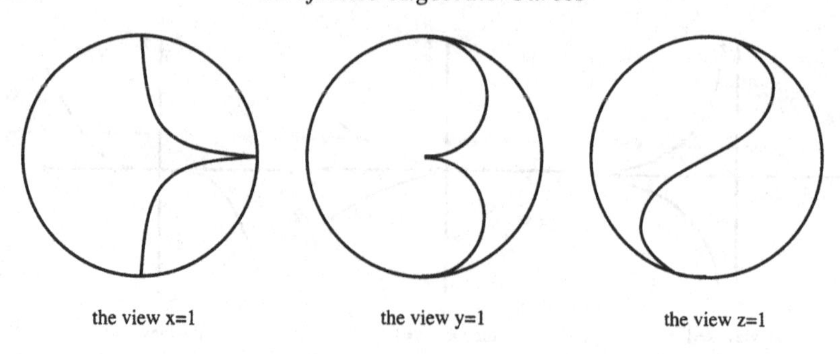

Fig. 9.8. Three hemispherical views of a cubic curve

$y = 0$, $z = 0$ in $\mathbb{R}^3$, we obtain the three pictures in Figure 9.8 which make it more plausible that they are different 'views' of the same curve F.

That brings us to one of the key constructions in this book, namely that of reversing the above process. More precisely, given any affine curve $f(x, y)$ of degree d in $\mathbb{K}^2$, and any affine plane L in $\mathbb{K}^3$ not through the origin, and not parallel to the z-axis, there exists a projective curve $F(x, y, z)$ of degree d in $P\mathbb{K}^2$ with the property that $F(x, y, z) = f(x, y)$ for every point (x, y, z) in L. Indeed, if L is defined by an equation $l = 1$, where $l = ax + by + cz$ with $c \neq 0$, and $f(x, y) = \sum a_{ij}x^i y^j$, we have only to observe that we can take

$$F(x, y, z) = l^d f\left(\frac{x}{l}, \frac{y}{l}\right) = \sum a_{ij}x^i y^j l^{d-i-j}.$$

Note that the condition $c \neq 0$ ensures that F has a term of degree d with non-zero coefficient. This is the process of *homogenization* with respect to the form l, you homogenize f by multiplying each monomial by that power of l necessary to bring the degree up to d. In practice, the only case of this construction we will meet is when L is the plane $z = 1$, so

$$F(x, y, z) = z^d f\left(\frac{x}{z}, \frac{y}{z}\right) = \sum a_{ij}x^i y^j z^{d-i-j}.$$

For instance, to the affine line $f = ax + by + c$ we associate the projective line $F = ax + by + cz$. Note that f itself is one of the family of affine views provided by F; indeed, if we homogenize f to obtain F, and then dehomogenize using z again, we are back to f. However, beware! That can fail when the order of the processes is reversed. For instance, if we dehomogenize the projective curve $F = z^2(x^2 + z^2)$, we obtain the affine curve $f = x^2 + 1$, and then homogenization with respect to z yields a *different* projective curve $G = x^2 + z^2$. Of course the process of homogenization applies equally well when the variables in the affine

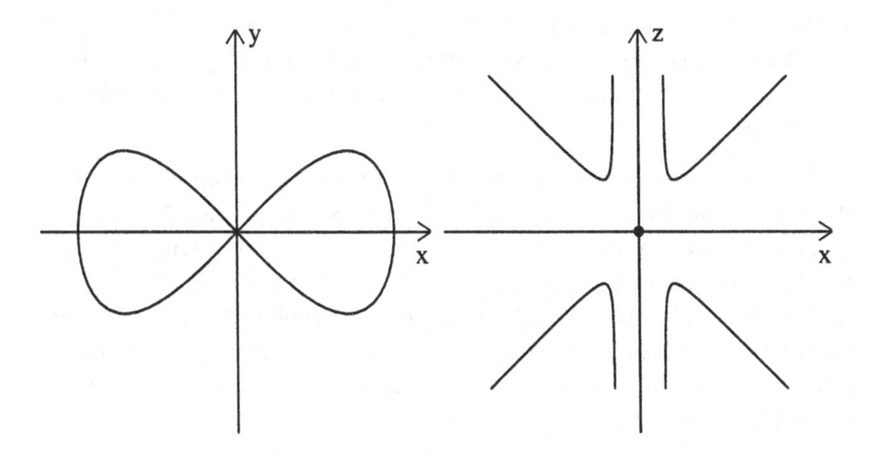

Fig. 9.9. Two affine views of the eight-curve

plane are labelled as x, z (when we homogenize with respect to y) or as y, z (when we homogenize with respect to x).

Example 9.13 Consider the cubic $f = y - x^3$ in $\mathbb{R}^2$. The associated projective curve is $F = yz^2 - x^3$. But, as we saw in Example 9.12, the zero sets of the principal affine views $f = y - x^3$, $g = z^2 - x^3$ and $h = yz^2 - 1$ are visually quite different. At first sight it is difficult to imagine that the views g, h have any connexion with the original curve f. In particular, the view g shows that there is a 'cusp' at infinity on f, something that one might not have expected.

This example makes a very fundamental conceptual point. An affine curve f is simply one out of a very large family of affine views of the associated projective curve F. By working solely in the affine case, we deny ourselves the rich geometry provided by the projective situation. It is virtually impossible to get a genuine understanding of the geometry of an affine curve without moving into the projective situation. Here is another illuminating illustration.

Example 9.14 The projective curve in $P\mathbb{R}^2$ associated to the eight-curve $f = y^2 - x^2 + x^4$ is $F = y^2z^2 - x^2z^2 + x^4$. This meets the line at infinity $z = 0$ when $x^4 = 0$, yielding the point at infinity $Y = (0 : 1 : 0)$ on the y-axis. The affine view given by $y = 1$ is $g = z^2 - x^2z^2 + x^4$, and in that view Y corresponds to the origin in the (z, x)-plane. The zero set of g is easily sketched, by writing $g = 0$ in the form $z^2 = x^4/(x^2-1)$, and reveals that the origin is an isolated point on the zero set of g. (Figure 9.9.) Thus Y is an 'isolated' point on the projective curve – another unexpected revelation.

There is a very simple relationship between the zero sets of an affine curve f and its homogenization F with respect to z. Since $F(x, y, 1) = f(x, y)$ we have $f(x, y) = 0$ if and only if $F(x, y, 1) = 0$. Identifying the point (x, y) in $\mathbb{K}^2$ with the point $(x : y : 1)$ in $P\mathbb{K}^2$ (as above), we see that the zero set of F can be thought of as the zero set of f, plus the points at infinity $(x : y : 0)$ satisfying $F(x, y, 0) = 0$. Note that $F(x, y, 0)$ is a binary form in x, y, so finding the points at infinity on the curve is tantamount to finding the linear factors of that form.

Example 9.15 Any projective curve F in $P\mathbb{C}^2$ has an infinite zero set. Indeed the zero set of F 'contains' the zero set of any affine curve f in $\mathbb{C}^2$ obtained by dehomogenization, so is infinite by Lemma 2.1.

Example 9.16 To any line $ax + by + c$ in $\mathbb{K}^2$ we can associate a line $ax + by + cz$ in $P\mathbb{K}^2$. The zero set of the projective line comprises that of the affine line together with a unique point at infinity $(x : y : 0)$ for which $ax + by = 0$, i.e. the point $(-b : a : 0)$. Note that the point at infinity on the line does not depend on the coefficient c. All lines parallel to $ax+by+c = 0$ in $\mathbb{K}^2$ have the same point at infinity. In particular, the point at infinity on the x-axis is $X = (1 : 0 : 0)$ and the point at infinity on the y-axis is $Y = (0 : 1 : 0)$.

Example 9.17 The parabola $f = y^2 - x$ in $\mathbb{R}^2$ gives rise to the conic $F = y^2 - xz$ in $P\mathbb{R}^2$ which meets the line at infinity $z = 0$ when $F(x, y, 0) = y^2 = 0$, i.e. at the single point $X = (1 : 0 : 0)$.

Example 9.18 The Maltese cross (Figure 6.6) is the quartic in $\mathbb{R}^2$ defined by $f = xy(x^2 - y^2) - (x^2 + y^2)$. The associated projective curve in $P\mathbb{R}^2$ is $F = xy(x^2 - y^2) - z^2(x^2 + y^2)$, which meets the line at infinity $z = 0$ when $xy(x^2 - y^2) = 0$. Thus we obtain the points at infinity on the lines $x = 0$, $y = 0$, $y = x$ and $y = -x$, namely $(0 : 1 : 0)$, $(1 : 0 : 0)$, $(1 : 1 : 0)$ and $(1 : -1 : 0)$.

Example 9.19 The general circle in $\mathbb{R}^2$ with centre (a, b) and radius $r > 0$ is $f = (x - a)^2 + (y - b)^2 - r^2$. The associated projective curve is $F = (x - az)^2 + (y - bz)^2 - r^2z^2$ in $P\mathbb{R}^2$. This meets the line at infinity $z = 0$

when $x^2 + y^2 = 0$, giving no points in $\mathbb{PR}^2$. However, we can also think of F as a projective curve in $\mathbb{PC}^2$, in that case the relation $x^2 + y^2 = 0$ yields $y = \pm ix$, so there are two points at infinity in $\mathbb{PC}^2$, namely the complex conjugate points $I = (1 : i : 0)$ and $J = (1 : -i : 0)$.

The points I, J in Example 9.19 play a special role in the history of curve theory; they are called the *circular points at infinity*, since any circle (thought of as a complex projective curve) passes through them. The notation has its origins in the biblical names Isaak and Jakob. Real algebraic curves for which the associated complex projective curve passes through I, J are said to be *circular*, and play an important role in mechanical engineering problems. We have already met a number of circular curves, for instance, the pedal curves of the parabola and the unit circle, the three- and four-leaved clovers, and the astroid.

Exercises

9.5.1 Find the points at infinity on the projective curves in $\mathbb{PR}^2$ corresponding to the following curves f in $\mathbb{R}^2$.

(i) $f = x^3 - xy^2 - y$
(ii) $f = (y - x^2)^2 - xy^3$
(iii) $f = x^4 + 4xy - y^4$
(iv) $f = x^5 + y^5 - 2xy^2$
(v) $f = x^2y^2 + x^2 - y^2$.

9.5.2 Find the points at infinity on the projective curves in $\mathbb{PR}^2$ corresponding to the following curves f in $\mathbb{R}^2$.

(i) $f = x^5 + y^5 - 2x^2 - 5xy + 2y^2$
(ii) $f = x^5 + y^5 - 5x^2y^2 + 3x - 3y$
(iii) $f = (2x + y)(x + y)^2 + (2x - y)(x + y) + (x - y)$
(iv) $f = x^4y + x^2y^3 + x^4 - y^3 + 1$.

9.5.3 Let $f = cf_1^{r_1} \dots f_s^{r_s}$ be the unique factorization of the polynomial f into irreducible factors, and let $F, F_1, \dots, F_s$ be the forms obtained from $f, f_1, \dots, f_s$ by homogenization. Show that $F = cF_1^{r_1} \dots F_s^{r_s}$ is the unique factorization of the form F into irreducible forms.

9.5.4 Let h be any conic in $\mathbb{R}^2$ having at least one real point and let H be the corresponding projective conic in $\mathbb{PC}^2$. Show that if H passes through the circular points at infinity I, J then

h is a circle. (Point out *precisely* where you use the hypothesis that h is irreducible, and where you use the hypothesis that h has a real point.)

9.5.5 Let $f(x, y)$ be a curve of degree d in $\mathbb{R}^2$, and let $F(x, y, z)$ be the corresponding projective curve in $P\mathbb{C}^2$. Show that $f(x, y)$ is circular if and only if its HOT are divisible by $x^2 + y^2$. Use this characterization to verify that the pedal curves of the parabola and the unit circle, the three- and four-leaved clovers, and the astroid are all circular curves.

9.5.6 Let F, G be two curves in $P\mathbb{R}^2$ intersecting in finitely many points. Show that the intersections are either real, or occur in complex conjugate pairs.

10

Singularities of Projective Curves

The next step in our development is to extend the concept of 'singularity' from the affine to the projective situation. We will follow the same pattern as in the affine case. The starting point is to extend the idea of intersection number (of a line with a curve at some point) to the projective situation; we can then define the 'multiplicity' of a point on a projective curve, and use this to introduce 'singular' points. As an application of these ideas, we return to conics, and show that by thinking of conics in the projective (rather than the affine) plane we can gain geometric insight into the delta invariants of Chapter 5.

10.1 Intersection Numbers

For projective curves we can mimic what we did in Chapter 6 for affine curves. Let $F(x, y, z)$ be a curve of degree d in $\mathbb{PK}^2$. And let L be a line. L is determined by any two distinct points, say $A = (a_1 : a_2 : a_3)$ and $B = (b_1 : b_2 : b_3)$: then, as in Example 9.5, the line L can be parametrized as $x_1 = sa_1 + tb_1$, $x_2 = sa_2 + tb_2$, $x_3 = sa_3 + tb_3$. The intersections of F with L are given by $\Phi(s,t) = 0$, where Φ is defined by

$$\Phi(s,t) = F(sa_1 + tb_1, sa_2 + tb_2, sa_3 + tb_3).$$

Since $F(x, y, z)$ is a form of degree d we have $\Phi(\lambda s, \lambda t) = \lambda^d \Phi(s,t)$ for any scalar λ, so $\Phi(s,t)$ is a binary form of degree d, the *intersection form*. Of course, Φ may be identically zero. It is worth comparing this with the situation in the affine case where an intersection polynomial $\phi(t)$, associated to an affine curve f of degree d and a line l, is only known to be of degree $\leq d$. Concerning the generalities of how lines intersect projective curves, we have the following analogue of Lemma 4.4.

Lemma 10.1 *Let F be a curve of degree d in $\mathbb{PK}^2$, and let L be a line. Then either L meets F in $\leq d$ points, or L is a component of F. Thus, for a curve F with no line components, the degree d is an upper bound for the number of intersections of F with any line L.*

Proof As above, we can associate an intersection form Φ of degree d to the (parametrized) line L and the curve F. Provided Φ is not identically zero, the equation $\Phi(s,t) = 0$ has $\leq d$ roots, and there are $\leq d$ points of intersection. Suppose now that Φ is identically zero, so F vanishes at every point on L. Let L have equation $ax + by + cz = 0$. It is no restriction to suppose that $c \neq 0$, so $X = x$, $Y = y$, $Z = ax + by + cz$ is an invertible linear mapping. Then $F(X,Y,Z)$ vanishes at every point of $Z = 0$, so the binary form $F(X,Y,0)$ is identically zero, and Z must be a factor of $F(X,Y,Z)$. In terms of the variables x, y, z that means that L is a factor of $F(x,y,z)$, as required. □

Next, let us look in more detail at the way in which a line L intersects a projective curve F of degree d. Assume L is not a component of F. Let P be a point on L corresponding to the ratio $(s_0 : t_0)$, in some parametrization of L. We define the *intersection number* $I(P,F,L)$ to be the multiplicity of $(s_0 : t_0)$ as a root of the equation $\Phi(s,t) = 0$; that means that $\Phi(s,t)$ has a factor $(t_0 s - s_0 t)^m$ for some integer $m \geq 0$, and that $I(P,F,L) = m$. Clearly $I(P,F,L)$ is an integer ≥ 0, and vanishes if and only if P is not an intersection of F, L. It is a convention that $I(P,F,L) = \infty$ when L is a component of F.

Example 10.1 The projective curve corresponding to the eight-curve $f = y^2 - x^2 + x^4$ in $\mathbb{R}^2$ is $F = y^2z^2 - x^2z^2 + x^4$ in $\mathbb{PR}^2$. Let L be the line $x = 0$. Then L is the line through $P = (0:0:1)$, $Q = (0:1:0)$ and parametrized as $x = 0$, $y = t$, $z = s$. The resulting intersection form is $\Phi(s,t) = s^2t^2$, having roots $(1:0)$ and $(0:1)$ both of multiplicity 2; the root $(1:0)$ corresponds to P, whilst $(0:1)$ corresponds to Q. Since both roots have multiplicity 2 the intersection numbers are $I(P,F,L) = 2$, $I(Q,F,L) = 2$.

The next step in our development is to establish that $I(P,F,L)$ does not depend on the chosen parametrization of the line L. We follow the paradigm of the affine case, by establishing a preliminary lemma. By a projective *change of parameter* we mean an invertible linear map $U(s,t) = (u,v)$ where $u = \alpha s + \gamma t$, $v = \beta s + \delta t$ and α, β, γ, δ are scalars with $\alpha\delta - \beta\gamma \neq 0$. We now have

Lemma 10.2 *Suppose that two forms $\Phi(s,t)$, $\Psi(u,v)$ are related by a projective change of parameter, i.e. $\Phi(s,t) = \Psi(U(s,t))$ for some projective change of parameter $U(s,t)$. Then (s_0,t_0) is a zero of multiplicity m of $\Phi(s,t)$ if and only if $U(s_0,t_0)$ is a zero of multiplicity m of $\Psi(u,v)$.*

Proof Write $U(s_0,t_0) = (u_0,v_0)$. Observe (using the Factor Theorem) that $-t_0s+s_0t$ is a factor of $\Phi(t)$ if and only if (s_0,t_0) is a zero of $\Phi(t)$, if and only if $U(s_0,t_0)$ is a zero of $\Psi(U(s,t))$, if and only if $-v_0u+u_0v$ is a factor of $\Psi(u,v)$. However, by Lemma 4.6 corresponding factors $t-t_0$, $u-u(t_0)$ have the same multiplicity, as required. □

Lemma 10.3 *$I(P,F,L)$ depends only on P, F, L, it does not depend on the choice of parametrization of L.*

Proof This follows exactly the same lines as the corresponding result (Lemma 6.3) in the affine case. Consider two parametrizations of L, the first via distinct points A, B with parameter the ratio $(s:t)$, yielding the intersection form $\Phi(s,t) = F(sA+tB)$, and the second via distinct points C, D with parameter the ratio $(u:v)$, yielding the intersection form $\Psi(u,v) = F(uC+vD)$. The connexion between $\Phi(s,t)$, $\Psi(u,v)$ is easily established. Let $A = \alpha C + \beta D$, $B = \gamma C + \delta D$. Then, by direct calculation

$$\Phi(s,t) = F(sA+tB) = F(u(s,t)C + v(s,t)D) = \Psi(u(s,t), v(s,t))$$

where $u(s,t) = \alpha s + \gamma t$, $v(s,t) = \beta s + \delta t$. Thus $\Phi(s,t)$ is obtained from $\Psi(u,v)$ by the projective change of parameter $U(s,t) = (u(s,t), v(s,t))$. (Note that $\alpha\delta - \beta\gamma \neq 0$ else A, B coincide.) Then $(s_0:t_0)$ is a zero of $\Phi(s,t)$ if and only if $U(s_0,t_0)$ is a zero of $\Psi(u,v)$, and it follows from Lemma 10.2 that the multiplicities of the zeros coincide. The result follows from the definition of $I(P,F,L)$. □

The *total number* of intersections of F with L (counted properly) is the sum of the intersection numbers $I(P,F,L)$ with P on F. As the sum of the multiplicities of the roots of $\Phi(s,t)$ is $\leq d$, we deduce that the total number of intersections of a curve F of degree d with a line L is $\leq d$. Over the complex field the intersection form Φ has exactly d zeros, counted with multiplicities, so the total number of intersections (counted properly) is *precisely* d, in particular Φ has at least one zero. Thus we have established

Lemma 10.4 *Let F be a curve of degree d in $\mathbb{PC}^2$, and let L be a line which is not a component of F. Then L meets F in exactly d points, counted properly. In particular, L meets F in at least one point.*

Lemma 10.4 is a special case of a celebrated result in the theory of algebraic curves, namely Bézout's Theorem. Roughly speaking, this says the following. Any two curves F, G in $\mathbb{PC}^2$ with no common component intersect in only finitely many points P. Moreover, it is possible to associate an 'intersection number' $I(P,F,G)$ to any such curves F, G, and any point P, in such a way that the sum of the intersection numbers at all points of intersection is de, where $d = \deg F$, $e = \deg G$. And in the special case when G is a line L, this 'intersection number' agrees with the number $I(P,F,L)$ defined above. The proof of Bézout's Theorem is presented in Chapter 14, together with the algebraic ideas on which it depends.

It is worth remarking at this stage that the definition of $I(P,F,L)$ has a fatal limitation. We used the fact that L can be parametrized globally by parameters s, t. In principle, one could extend the idea by replacing L by any curve G which can be globally parametrized. However, this approach founders on the fundamental fact that rather few algebraic curves have this property. (We will have more to say about this matter in the final chapter.) In practice you can reduce the computation of intersection numbers in the projective plane to one in an affine plane, via the following result.

Lemma 10.5 *Let P be an intersection point of a curve F of degree d with a line L in the projective plane $\mathbb{PK}^2$. Dehomogenize with respect to one of the variables x, y, z to obtain a point p, a curve f, and a line l in the affine plane $\mathbb{K}^2$. Then $I(P,F,L) = I(p,f,l)$.*

Proof At least one of the homogeneous coordinates of P is non-zero. Without loss of generality, assume it is z. Let $B = (b_1 : b_2 : 0)$ be the unique point where $z = 0$ meets L, and let $A = (a_1 : a_2 : 1)$ be any point on L distinct from B. Then the projective line L is parametrized as $sA + tB$, with $s = 0$ corresponding to B, and gives rise to an intersection form $\Phi(s,t) = F(sA + tB)$. Likewise, the affine line l is parametrized as $a + tb$, where $a = (a_1, a_2)$, $b = (b_1, b_2)$, and gives rise to an intersection polynomial $\phi(t) = f(a + tb)$. Moreover, under the identification $(x, y) \sim (x : y : 1)$ the point on l with parameter t_0 corresponds to the point on

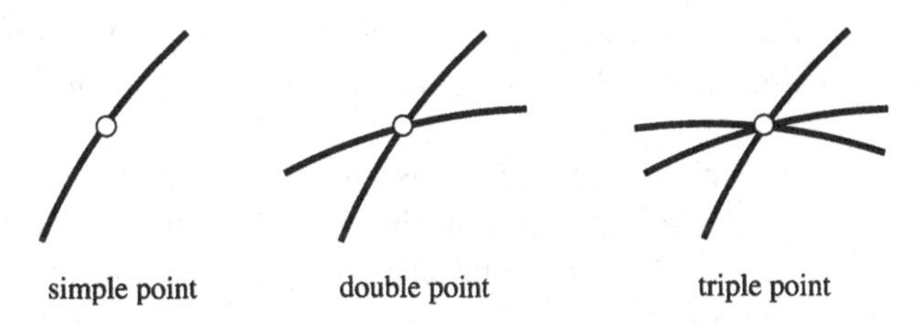

Fig. 10.1. Simple, double and triple points

L with parameter $s = 1$, $t = t_0$. For $s \neq 0$ we have

$$\Phi(s,t) = F(sA + tB) = s^d f(a + ub) = s^d \phi(u)$$

where $u = t/s$. Thus the zeros (s_0, t_0) of $\Phi(s,t)$ with $s_0 \neq 0$ correspond to zeros t_0 of $\phi(t)$, and have the same multiplicity. The result follows from the definitions of $I(P,F,L)$, $I(p,f,l)$. □

Example 10.2 Consider the curve $F = y^2z^2 - x^2z^2 + x^4$ in $\mathbb{PR}^2$, the line L given by $x = 0$, and the point $P = (0:0:1)$. Dehomogenizing with respect to z, we obtain the affine curve $f = y^2 - x^2 + x^4$, the line l given by $x = 0$, and the point $p = (0,0)$. We can parametrize l as $x = 0$, $y = t$ with p corresponding to the value $t = 0$ of the parameter. Then the intersection polynomial is $\phi(t) = t^2$, and $I(p,f,l) = 2$, agreeing with the result of Example 10.1.

10.2 Multiplicity of a Point on a Curve

Let P be a point on a curve F in $\mathbb{PK}^2$. Consider the pencil of all lines L through P. By the *multiplicity* of P on F we mean the minimal value m of the intersection number $I(P,F,L)$ as L varies over the pencil, i.e. $I(P,F,L) \geq m$ for *every* line L through P, and there exist lines L for which $I(P,F,L) = m$. Points of multiplicity 1, 2, 3, 4, ... are said to be *simple*, *double*, *triple*, *quadruple*, ... points of F. A crude mental picture for a point P of multiplicity m is that there are m 'branches' of the curve self-intersecting at P. (Figure 10.1.) But, as in the affine case, that is not the whole story.

In view of Lemma 10.5, we have the following result, which allows us to compute multiplicities via the technique of Lemma 6.4.

Lemma 10.6 *Let P be a point on a curve F in $\mathbb{PK}^2$. Dehomogenize with* *respect to one of the variables x, y, z to obtain a point p on a curve f in $\mathbb{K}^2$. Then the multiplicity of P on F coincides with that of p on f.*

Example 10.3 The eight-curve gives rise to the projective curve $F = y^2z^2 - x^2(z^2 - x^2)$ in $\mathbb{PR}^2$, intersecting the line at infinity at the unique point $Q = (0 : 1 : 0)$. To find the multiplicity of Q we dehomogenize with respect to y to obtain the affine curve $f = z^2 - x^2(z^2 - x^2)$ and the point $q = (0, 0)$ in the (x, z)-plane. Here the LOT are z^2, so q has multiplicity 2 on f, and hence Q has multiplicity 2 on F.

Exercises

10.2.1 Let F be an irreducible quartic curve in $\mathbb{PC}^2$. Show that F cannot have a point P of multiplicity ≥ 4. Show also that if F has a triple point P then P is the only point on F of multiplicity ≥ 2.

10.2.2 Let P be a point on a projective curve F in $\mathbb{PK}^2$ of multiplicity $\geq m$. Show that P is a point on the projective curve F_z of multiplicity $\geq (m - 1)$. And show, by means of an example, that P could have multiplicity $> m - 1$.

10.3 Singular Points

A point P on a projective curve F is *singular* when it has multiplicity ≥ 2. The curve F itself is *singular* when it has at least one singular point, otherwise it is *non-singular*. We will see in Chapter 18 that an irreducible curve F has at most finitely many singular points.

Example 10.4 A cubic curve F in $\mathbb{PK}^2$ with two singular points P, Q necessarily reduces to a conic, and the line L joining P, Q. Indeed, since P, Q are singular, we have $I(P, F, L) \geq 2$, $I(Q, F, L) \geq 2$ so the total number of intersections of L with F is ≥ 4, contradicting Lemma 10.1 unless L is a component of F.

Here is a simple criterion for a point to be singular on a curve F, analogous to Lemma 6.6 in the affine case.

Lemma 10.7 *A point P is a singular point on a curve F if and only if $F(P) = 0$ and $F_x(P) = F_y(P) = F_z(P) = 0$.*

Proof The condition for $P = (p_1 : p_2 : p_3)$ to be singular is that $I(P, F, L) \geq 2$ for *every* line L through P. We need to interpret this in detail. Let L be any line through P, and let $Q = (q_1 : q_2 : q_3)$ be another point on L. Recall that the intersection form $\Phi(s,t) = F(sP + tQ)$ is a form of degree d in s, t so has the form

$$\Phi(s,t) = s^d F_0(P,Q) + s^{d-1} t F_1(P,Q) + \cdots + t^d F_d(P,Q)$$

where each $F_k(P,Q)$ is a polynomial in the entries of P, Q. The point P corresponds to the ratio $(1 : 0)$ and the condition $I(P,F,L) \geq 2$ is that $(1 : 0)$ should have multiplicity ≥ 2 as a root of $\Phi(s,t) = 0$, i.e. that $\Phi(s,t)$ has t^2 as a factor, i.e. that $F_0(P,Q) = 0$, $F_1(P,Q) = 0$. The condition $F_0(P,Q) = 0$ holds automatically, as P lies on F, so it remains to interpret the condition $F_1(P,Q) = 0$. Write

$$\phi(t) = F(P + tQ) = tF_1(P,Q) + \cdots + t^d F_d(P,Q)$$

so $F_1(P,Q) = \phi'(0)$. The Chain Rule gives $\phi'(t) = q_1 F_x + q_2 F_y + q_3 F_z$, where the partials are evaluated at $P + tQ$; setting $t = 0$ yields

$$F_1(P,Q) = q_1 F_x(P) + q_2 F_y(P) + q_3 F_z(P).$$

This expression is zero for all choices of q_1, q_2, q_3 if and only if $F_x(P) = F_y(P) = F_z(P) = 0$, as required. □

Example 10.5 A point R of intersection of two projective curves F, G is necessarily a singular point of $H = FG$. The point R satisfies the relations $F(R) = 0$, $G(R) = 0$ and hence $H(R) = 0$. Moreover, the rule for differentiating a product gives

$$H_x = F_x G + F G_x, \quad H_y = F_y G + F G_y, \quad H_z = F_z G + F G_z$$

so R also satisfies the relations $H_x(R) = 0$, $H_y(R) = 0$, $H_z(R) = 0$ and is therefore singular on H.

The labour involved in finding singular points in practice can be reduced via the following useful little result about forms, known as the Euler Lemma.

Lemma 10.8 *Let $F(x_1, \ldots, x_n)$ be a form of degree d over a field $\mathbb{K}$, then we have the Euler relation*

$$dF = \sum x_k F_{x_k}.$$

Proof Recall first (Lemma 3.12) that a form $F(x_1, \ldots, x_n)$ of degree d satisfies the following identity, for all $t \in \mathbb{K}$

$$F(tx_1, \ldots, tx_n) = t^d F(x_1, \ldots, x_n).$$

Differentiating both sides of this relation with respect to t, using the Chain Rule, we get the following relation. The required result then follows on setting $t = 1$.

$$\sum x_k F_{x_k}(tx_1, \ldots, tx_n) = dt^{d-1} F(tx_1, \ldots, tx_n).$$

□

The point of the Euler Lemma is that if all the partials F_x, F_y, F_z of a projective curve F vanish at a point P then *automatically* F itself vanishes at P. Thus *the singular points of a projective curve F are obtained by solving the equations*

$$0 = F_x = F_y = F_z.$$

Many of the classical affine curves turn out to have singularities 'at infinity', helping to explain their geometry.

Example 10.6 Agnesi's versiera (Example 2.8) is the real affine cubic $f = x^2y - 4a^2(2a - y)$ with $a > 0$. Although it has no singularities in $\mathbb{R}^2$, the corresponding projective curve $F = x^2y - 4a^2z^2(2az - y)$ does have a singularity at infinity. Singularities are given by $0 = F_x = 2xy$, $0 = F_y = x^2 + 4a^2z^2$, $0 = F_z = -8a^2z(3az - y)$. The first relation gives $x = 0$ or $y = 0$; $x = 0$ leads to the singularity $P = (0 : 1 : 0)$, the point at infinity on the y-axis, whilst $y = 0$ produces only the trivial solution. The nature of the singularity at P is revealed by setting $y = 1$ in F, yielding the affine curve

$$g = x^2 - 4a^2z^2(2az - 1) = (x^2 + 4a^2z^2) + \text{HOT}$$

having a singularity at the origin in the (x, z)-plane, namely an acnode. In particular, this explains why all the 'vertical' lines in the (x, y)-plane only meet the curve once – they meet the curve twice more at the point P.

Example 10.7 The eight-curve gives rise to the projective curve $F = y^2z^2 - x^2(z^2 - x^2)$ in $P\mathbb{R}^2$. Singular points require $0 = F_x = 2x(2x^2 - z^2)$, $0 = F_y = 2yz^2$, $0 = F_z = 2z(y^2 - x^2)$. Clearly, $y = 0$ or $z = 0$: $y = 0$ gives $P = (0 : 0 : 1)$ and $z = 0$ gives $Q = (0 : 1 : 0)$. This throws light on Example 10.1; there we saw that the line $x = 0$ meets the curve solely at P, Q with intersection number 2 in each case. Since Q is a singular point of the curve, that is no longer too surprising.

Example 10.8 Consider the family of *Steiner* cubics in $\mathbb{PC}^2$ given by the formulas

$$F = \mu(x^3 + y^3 + z^3) + 3\lambda xyz.$$

We claim that F is singular if and only if $\mu = 0$ or $\lambda^3 = -1$, and that in each of these four cases F reduces to a triangle. (By a *triangle* we mean a projective cubic which reduces to three non-concurrent lines, known as the *sides*; the three points of intersection of the sides are the *vertices* of the triangle.) When $\mu = 0$, the cubic is $F = xyz$, a triangle with singular points at the vertices $(1 : 0 : 0)$, $(0 : 1 : 0)$, $(0 : 0 : 1)$. When $\mu \neq 0$ we can assume that $\mu = 1$, by multiplying the equation through by a scalar. Then singular points are given by $F_x = 0$, $F_y = 0$, $F_z = 0$, i.e. by

$$x^2 = -\lambda yz, \qquad y^2 = -\lambda zx, \qquad z^2 = -\lambda xy.$$

Note that if one of x, y, z is $= 0$ then all are $= 0$, so we can assume x, y, z are all $\neq 0$. Equating the product of the three left-hand sides, with the product of the three right-hand sides, we get $(1 + \lambda^3)x^2y^2z^2 = 0$. When $\lambda^3 \neq -1$, one of x, y, z must be $= 0$, a contradiction, so in that case F is non-singular. Suppose $\lambda^3 = -1$. Multiplying the three relations by x, y, z respectively yields $x^3 = y^3 = z^3 = -\lambda xyz$. Let 1, ω, ω^2 be the three complex cube roots of 1. Then λ is one of -1, $-\omega$, $-\omega^2$; moreover it is no restriction to suppose that $x = 1$ so y, z must each be one of 1, ω, ω^2. That yields the following singular points.

$$\begin{cases} \lambda = -1 & (1 : 1 : 1) \quad (1 : \omega : \omega^2) \quad (1 : \omega^2 : \omega) \\ \lambda = -\omega & (1 : 1 : \omega^2) \quad (1 : \omega : \omega) \quad (1 : \omega^2 : 1) \\ \lambda = -\omega^2 & (1 : 1 : \omega) \quad (1 : \omega : 1) \quad (1 : \omega^2 : \omega^2). \end{cases}$$

In each case, we have three non-collinear points; the lines joining any two of these points meet F in ≥ 4 points, so must be components, and the cubic reduces to a triangle, with vertices the given singular points.

Exercises

10.3.1 Find the singular points of the quartic curve in $\mathbb{PC}^2$ given by $F = x^2y^2 + y^2z^2 + z^2x^2$, and determine their multiplicities.

10.3.2 Find the singular points of the quartic curve in $\mathbb{PC}^2$ given by $F = (x^2 - z^2)^2 - y^2z(2y + 3z)$, and determine their multiplicities.

10.3.3 Show that the quartic curve $F = (x^2 + y^2 - zx)^2 - z^2(x^2 + y^2)$ in $\mathbb{PC}^2$ has exactly three singular points.

10.3.4 Let λ be a complex number and let F_λ be the quartic curve in $\mathbb{PC}^2$ given by $F_\lambda = x^4 + xy^3 + y^4 - \lambda zy^3 - 2x^2yz - xy^2z + y^2z^2$. Show that $O = (0:0:1)$ is singular on F_λ. Further, show that there are exactly two values of λ for which F_λ has a second singular point.

10.3.5 Show that the quintic $F = (x^2 - a^2z^2)^2y - (y^2 - b^2z^2)^2x$ in $\mathbb{PC}^2$ has exactly four singular points. ($a \neq 0$, $b \neq 0$.)

10.3.6 Show that the sextic curve $F = (z^2 - x^2 - y^2)^3 - 27x^2y^2z^2$ in $\mathbb{PC}^2$ has exactly ten singular points. (This is the complex projective version of the astroid.)

10.3.7 Show that the sextic curve $F = (x^2 + y^2)^3 - 4x^2y^2z^2$ in $\mathbb{PC}^2$ has exactly three singular points, namely a quadruple point P, and two double points at the circular points at infinity I, J.

10.3.8 The Euler Lemma expresses a form F of degree d in terms of the first order partials of F. Extend this to second order partials by showing that

$$d(d-1)F = \sum x_i x_j F_{x_i x_j}.$$

10.3.9 Let $f(x, y)$ be a polynomial over the field $\mathbb{K}$ which can be written in the form $f = F + G$ where F, G are forms of different degrees $d, e \geq 2$. Use the Euler Lemma to show that the singular points of f in $\mathbb{K}^2$ satisfy the relations $F = 0$, $G = 0$. If, in addition, F, G have no common factors, deduce that $(0, 0)$ is the only singularity of f. Deduce that the Maltese cross curve in $\mathbb{R}^2$ defined by $f = xy(x^2 - y^2) - (x^2 + y^2)$ has just one singularity, namely a double point at $(0, 0)$.

10.3.10 Let $p(x)$ be a complex polynomial of degree $d \geq 3$, let f be the affine curve in $\mathbb{C}^2$ defined by $f = y^2 - p(x)$, and let F be the corresponding projective curve in $\mathbb{PC}^2$. Show that F meets the line at infinity at a unique point P, of multiplicity $(d-2)$. (The singular points of f were determined in Exercise 6.3.9.)

10.4 Delta Invariants viewed Projectively

In this section we will consolidate our account of affine conics in Chapter 5 by showing that the projective viewpoint gives considerable geometric insight into the delta invariants δ, Δ of that chapter. Recall that a general conic in $\mathbb{K}^2$ has the shape

$$q = ax^2 + 2hxy + by^2 + 2gx + 2fy + c \qquad (\star)$$

where the coefficients a, b, c, f, g, h lie in $\mathbb{K}$, and at least one of a, b, h is non-zero. The corresponding projective curve in $P\mathbb{K}^2$ is

$$Q = ax^2 + by^2 + cz^2 + 2hxy + 2fyz + 2gzx \qquad (**)$$

The first thing to ask is how the conic meets the line at infinity. Setting $z = 0$ in (**) gives a binary quadratic equation $ax^2 + 2hxy + by^2 = 0$. Over the complex field this has two roots (giving two points at infinity on the conic) with coincidence if and only if the discriminant vanishes. The discriminant is $-4(ab - h^2) = -4\delta$, where δ is the affine invariant we met in Chapter 5, thus the geometric significance of the relation $\delta = 0$ is that the associated projective conic meets the line at infinity at a single point with intersection number 2. Over the real field we can say more. When $\delta < 0$, the binary quadratic has two real roots (and the conic has two real points at infinity), and when $\delta > 0$ the binary quadratic has no real roots (and the conic has no real points at infinity). Thus in the real case, the conditions $\delta < 0$, $\delta = 0$, $\delta > 0$ correspond to the projective conic Q having two, one or no points at infinity. Looking at Table 5.4 we see that for an *irreducible* real conic these conditions characterize respectively the hyperbola, the parabola and the two types of ellipse.

A geometric understanding of the invariant Δ is based on the following elementary result.

Lemma 10.9 *The conic (**) is singular if and only the matrix A below is singular.*

$$A = \begin{pmatrix} a & h & g \\ h & b & f \\ g & f & c \end{pmatrix}$$

Proof Q is singular if and only if there exists a non-trivial solution (x, y, z) for the following system of linear equations: and by linear algebra that is the case if and only if the matrix A of coefficients is singular.

$$\begin{aligned} Q_x &= 2(ax + hy + gz) = 0 \\ Q_y &= 2(hx + by + fz) = 0 \\ Q_z &= 2(gx + fy + cz) = 0. \end{aligned}$$

□

Recall from linear algebra that a matrix A is singular if and only if its determinant is zero. However the determinant of A is precisely the invariant Δ of Chapter 5, so the geometric interpretation of the condition $\Delta = 0$ is that the associated projective conic Q has a singular point.

Example 10.9 The conic $Q(x, y, z) = xy + yz + zx$ in $\mathbb{PK}^2$ is non-singular because its matrix A (below) has non-zero determinant.

$$A = \frac{1}{2}\begin{pmatrix} 0 & 1 & 1 \\ 1 & 0 & 1 \\ 1 & 1 & 0 \end{pmatrix}.$$

Lemma 10.10 *A reducible conic Q is necessarily singular. And a singular conic Q with at least two points is reducible. Over the complex field a conic is singular if and only if it is reducible.*

Proof Suppose Q reduces, necessarily, to two lines L, M. These lines intersect in at least one point R which is singular on Q, by Example 10.5. Conversely, suppose Q is singular, with at least two points. Let R be a singular point of Q, and let S be a point on Q distinct from R. Then the line joining R, S meets Q in ≥ 3 points, so by Lemma 10.1 is a component, and Q is reducible. The final statement is immediate from the fact that any curve in $\mathbb{PC}^2$ has an infinite zero set. □

Over the real field, a singular conic may fail to be reducible; for instance, although the conic $Q = x^2 + y^2$ in $\mathbb{PR}^2$ is irreducible, the zero set contains only one point $P = (0 : 0 : 1)$, which is singular on Q. Finally, it is instructive to look again at the lists of normal forms for real affine conics in Table 5.4. All the normal forms with $\Delta = 0$ do indeed exhibit a singular point in the projective plane, though in the 'parallel lines' cases that point is 'at infinity'.

11
Projective Equivalence

In this chapter, we continue the systematic process of extending our ideas from the affine situation to the projective, considering the natural relation of 'equivalence' for projective curves. We follow the underlying ideas of Chapter 4, replacing affine maps by their projective analogues, namely 'projective maps'. On this basis, we list projective conics over both the real and the complex fields, resulting in shorter (and in some respects more illuminating) lists than in the affine case.

11.1 Projective Maps

Let Φ be an invertible linear mapping of $\mathbb{K}^3$. By linear algebra, the image under Φ of any line through the origin is another. Thus Φ defines a (bijective) mapping $\tilde{\Phi}$ of $\mathbb{PK}^2$, called a *projective* map (or change of coordinates, or coordinate system). In practice Φ is given by a formula $\Phi(x, y, z) = (X, Y, Z)$ of the following form, where $A = (a_{ij})$ is an invertible 3×3 matrix with entries in $\mathbb{K}$.

$$\left\{ \begin{array}{lcl} X & = & a_{11}x + a_{12}y + a_{13}z \\ Y & = & a_{21}x + a_{22}y + a_{23}z \\ Z & = & a_{31}x + a_{32}y + a_{33}z. \end{array} \right.$$

The relation between the mappings Φ, $\tilde{\Phi}$ can be expressed as follows. Let $\pi : \mathbb{K}^3 - \{O\} \to \mathbb{PK}^2$ be the map $\pi(x, y, z) = (x : y : z)$, where $O = (0, 0, 0)$. Then $\tilde{\Phi}$ is defined by the property that $\tilde{\Phi} \circ \pi = \pi \circ \Phi$; in other words, for all non-zero vectors (x, y, z) we have

$$\tilde{\Phi}(x : y : z) = (X : Y : Z). \qquad (\star)$$

Example 11.1 Special types of projective maps are provided by *scalings*, of the form $X = ax$, $Y = by$, $Z = cz$ (with $a \neq 0$, $b \neq 0$, $c \neq 0$) and *coordinate switches* such as $X = y$, $Y = x$, $Z = z$.

The invertible linear maps of $\mathbb{K}^3$ form a group under the operation of composition, the *general linear group* $GL(3)$. It follows immediately from the general properties of invertible linear maps that projective maps of $P\mathbb{K}^2$ are bijections, and hence have inverses. One of the basic facts about projective maps is that they likewise form a group under the operation of composition, the *projective linear group* $PGL(2)$. We need to verify the group axioms of Section 2.2, the missing detail comprises Exercise 11.1.1.

- G1. $PGL(2)$ is closed under composition. Indeed if Φ, Ψ are two invertible linear maps of $\mathbb{K}^3$, giving rise to the projective maps $\tilde{\Phi}$, $\tilde{\Psi}$ of $P\mathbb{K}^2$, and $\Theta = \Psi \circ \Phi$ then $\tilde{\Theta} = \tilde{\Psi} \circ \tilde{\Phi}$ so the composite is projective.
- G2. Composition of maps is known to be an associative operation on the set of all maps from a given set to itself.
- G3. The identity map of $P\mathbb{K}^2$ is projective; it is the projective map arising from the identity linear map of $\mathbb{K}^3$.
- G4. $PGL(2)$ is closed under inversion. Indeed, if Φ is an invertible linear map of $\mathbb{K}^3$, with inverse Ψ, giving rise to the projective maps $\tilde{\Phi}$, $\tilde{\Psi}$ then $\tilde{\Psi} = \tilde{\Phi}^{-1}$, so the inverse is projective.

A useful tool for handling projective maps is the Four Point Lemma, below. Its proof depends on a simple lemma in linear algebra, requiring a preliminary definition. Four vectors in $\mathbb{K}^3$ are in *general position* when any three are linearly independent. For the proof of the lemma it will be convenient to write $e_1 = (1, 0, 0)$, $e_2 = (0, 1, 0)$, $e_3 = (0, 0, 1)$, $u = (1, 1, 1)$. Note that e_1, e_2, e_3, u are in general position.

Lemma 11.1 *Let p_1, p_2, p_3, p_4 be four vectors in $\mathbb{K}^3$ in general position. Then there exists an invertible linear map Φ taking p_1, p_2, p_3, p_4 in that order to non-zero scalar multiples of e_1, e_2, e_3, u. Moreover, Φ is unique, up to scalar multiples.*

Proof By the general position hypothesis p_1, p_2, p_3 are linearly independent, so $p_4 = ap_1 + bp_2 + cp_3$ for some scalars a, b, c, necessarily non-zero. And by linear algebra, there is an invertible linear map taking p_1, p_2, p_3 to e_1, e_2, e_3, and hence p_4 to (a, b, c). Composing this linear map with the scaling $(x, y, z) \to (x/a, y/b, z/c)$ produces a linear map with the desired properties. For uniqueness, suppose that Φ, Φ' are linear maps taking p_1, p_2, p_3, p_4 to αe_1, βe_2, γe_3, δu and $\alpha' e_1$, $\beta' e_2$, $\gamma' e_3$, $\delta' u$. Then $\delta u = \Phi(p_4) = (a\alpha, b\beta, c\gamma)$, $\delta' u = \Phi'(p_4) = (a\alpha', b\beta', c\gamma')$. It follows that (α, β, γ), $(\alpha', \beta', \gamma')$ are linearly dependent, implying that Φ, Φ' differ only by a scalar multiple. □

The immediate consequence of this result is that for two invertible linear maps Φ, Φ' we have $\tilde{\Phi} = \tilde{\Phi}'$ if and only $\Phi' = \lambda\Phi$ for some non-zero scalar λ. Thus a projective map of $P\mathbb{K}^2$ can be viewed as an invertible linear map of $\mathbb{K}^3$, *up to multiplication by non-zero scalars*.† The lemma translates into projective terms as follows. Four points in $P\mathbb{K}^2$ are said to be in *general position* when no three lie on a line; equivalently, any representatives of any three points are in general position in $\mathbb{K}^3$. The points E_1, E_2, E_3 in $P\mathbb{K}^2$ corresponding to e_1, e_2, e_3 are sometimes referred to as the *vertices of reference*, and the point U corresponding to u as the *unit point*. Note that E_1, E_2, E_3, U are in general position.

Lemma 11.2 *Let P_1, P_2, P_3, P_4 be four points in $P\mathbb{K}^2$ in general position. Then there exists a unique projective map $\tilde{\Phi}$ taking P_1, P_2, P_3, P_4 in that order to E_1, E_2, E_3, U.* (The Four Point Lemma.)

In fact the Four Point Lemma can be stated more generally. Given two sets P_1, P_2, P_3, P_4 and Q_1, Q_2, Q_3, Q_4 of four points in $P\mathbb{K}^2$ in general position, there exists a unique projective map taking P_1, P_2, P_3, P_4 in that order to Q_1, Q_2, Q_3, Q_4. That follows immediately from the above statement, and the fact that the projective mappings form a group.

Exercises

11.1.1 Let Φ, Ψ be invertible linear maps of $\mathbb{K}^3$, and let $\Theta = \Psi \circ \Phi$. Use the defining relation (⋆) to show that $\tilde{\Theta} = \tilde{\Psi} \circ \tilde{\Phi}$. Deduce that $\tilde{\Psi} = \tilde{\Phi}^{-1}$ when Ψ is the inverse of Φ.

11.1.2 Find the projective mapping of $P\mathbb{C}^2$ which maps the points $A = (3 : -2 : 1)$, $B = (-4 : 2 : -1)$, $C = (2 : -1 : 1)$, $D = (3 : 0 : 1)$ to $A' = (1 : 0 : 0)$, $B' = (0 : 1 : 0)$, $C' = (0 : 0 : 1)$, $D' = (1 : 1 : 1)$ respectively.

11.1.3 Find the projective mapping of $P\mathbb{C}^2$ which leaves the points $A = (2 : 0 : 1)$, $B = (-1 : \sqrt{3} : 1)$ and $C = (1 : \sqrt{3} : -1)$ fixed, and takes $D = (0 : 0 : 1)$ to $D' = (1 : 0 : 0)$.

† The reader familiar with the concept of a quotient group will see that in view of this remark one can describe the group $PGL(2)$ as the quotient of the general linear group $GL(3)$ by the subgroup comprising non-zero scalar multiples λI of the identity element I.

11.1.4 Let S, T, U, V be four lines in $\mathbb{PK}^2$ in *general position*, i.e. with the property that no three are concurrent. And let S', T', U', V' be another set of four lines in general position. Use the Four Point Lemma to show that there is a projective mapping of $\mathbb{PK}^2$ which takes S, T, U, V respectively to S', T', U', V'. (The Four Line Lemma.)

11.1.5 Show that the equation of any irreducible conic through $E_1 = (1:0:0)$, $E_2 = (0:1:0)$, $E_3 = (0:0:1)$ has the form $hxy + fyz + gzx = 0$ for some *non-zero* scalars f, g, h. By appropriate scaling of the variables, show that there is a projective map of $\mathbb{PK}^2$ which fixes E_1, E_2, E_3, and maps a given irreducible conic through them to the conic $xy + yz + zx$. For $i = 1, 2, 3$ let A_i (respectively B_i) be distinct points on an irreducible conic F (respectively G) in $\mathbb{PK}^2$. Show that there is a projective map of $\mathbb{PK}^2$ which takes A_i to B_i, for $i = 1, 2, 3$, and F to G. (Use the Four Point Lemma, and the previous part of the question.)

11.2 Projective Equivalence

The natural extension of the idea of 'equivalence' to projective curves is the following. Two projective curves F, G in $\mathbb{PK}^2$ are *projectively equivalent* when there exists an invertible linear mapping Φ of $\mathbb{K}^3$, and a scalar $\lambda \neq 0$, for which $G = \lambda(F \circ \Phi)$; more concretely, $G(x, y, z) = \lambda F(X, Y, Z)$ where

$$\left\{ \begin{array}{rcl} X &=& a_{11}x + a_{12}y + a_{13}z \\ Y &=& a_{21}x + a_{22}y + a_{23}z \\ Z &=& a_{31}x + a_{32}y + a_{33}z \end{array} \right.$$

and the matrix $A = (a_{ij})$ of coefficients is invertible. In this situation, the projective mapping $\tilde{\Phi}$ maps the zero set of G bijectively to the zero set of F. Note that projective equivalence is an equivalence relation, and that projectively equivalent curves F, G have the same degree: see Exercise 11.2.2. Moreover, F reduces if and only if G does. A possible objective is to classify curves of given degree up to projective equivalence. For curves of very low degree this can be achieved. The next example deals with lines, whilst conics are the object of Section 11.3.

Example 11.2 Any two lines L, L' in $\mathbb{PK}^2$ are projectively equivalent. It suffices to show that any line L is projectively equivalent to the line $x = 0$. The line L has an equation $ax + by + cz = 0$ for some scalars

a, b, c not all zero. By permuting the variables (a projective change of coordinates), we can suppose $a \neq 0$. Then $X = ax + by + cz$, $Y = y$, $Z = z$ is a projective change of coordinates (since the 3×3 matrix of coefficients has determinant $a \neq 0$) and in the new coordinates the line is given by $X = 0$. In particular, this example shows us that the zero sets of any two lines in $P\mathbb{K}^2$ are the 'same'. For instance by Example 9.6 the zero set of any line in $P\mathbb{R}^2$ can be thought of as a circle, and by Example 9.7 the zero set of any line in $P\mathbb{C}^2$ can be thought of as a sphere.

Here is a typical application of the Four Point Lemma, to the question of finding a 'prenormal' form for quartics in $P\mathbb{C}^2$ with three singular points.

Lemma 11.3 *Any irreducible quartic curve F in $P\mathbb{C}^2$ having three distinct double points A, B, C is projectively equivalent to a quartic of the following shape (with $abc \neq 0$) having double points at the vertices of reference*

$$F = ay^2z^2 + bz^2x^2 + cx^2y^2 + 2xyz(fx + gy + hz).$$

Proof Note first that A, B, C cannot be collinear, else the line on which they lie meets F in $\geq 2 + 2 + 2 = 6$ points, which is impossible by Lemma 10.1. By the Four Point Lemma we can assume that $A = (1 : 0 : 0)$, $B = (0 : 1 : 0)$, $C = (0 : 0 : 1)$. The condition for F to pass through these points is that the monomials x^4, y^4, z^4 do not occur in the defining polynomial for F. Now consider the condition for $A = (1 : 0 : 0)$ to be singular. This means that F_x, F_y, F_z all vanish at A, i.e. do not involve the monomial x^3. And that means that F cannot involve any of the monomials x^4, x^3y, x^3z having a factor x^3. Likewise the condition on F that $B = (0 : 1 : 0)$ is a singular point is that F cannot involve any of the monomials y^4, y^3x, y^3z having a factor y^3. And the condition for $C = (0 : 0 : 1)$ to be a singular point is that F doesn't involve any of the monomials z^4, z^3x, z^3y having a factor z^3. Thus F is a linear combination of the remaining six monomials of degree 4 in x, y, z, so has the required form. Finally, it is clear that a, b, c are all non-zero, since otherwise F reduces; for instance, if $a = 0$ then F has x as a factor. □

Exercises

11.2.1 Use the Four Line Lemma to show that any two lines in $P\mathbb{K}^2$ are projectively equivalent; likewise, show that any two line-pairs in $P\mathbb{K}^2$ are projectively equivalent.

Table 11.1. *The three types of complex projective conic*

normal form	type	rank
$x^2+y^2+z^2$	irreducible conic	3
x^2+y^2	line-pair	2
x^2	repeated line	1

11.2.2 Use the equivalent definition of homogeneity in Section 3.4 to show that projectively equivalent curves in $P\mathbb{K}^2$ have the same degree.

11.3 Projective Conics

Any conic in $P\mathbb{K}^2$ has the form $Q(x_1, x_2, x_3) = \sum a_{ij}x_ix_j$ with coefficients a_{ij} in $\mathbb{K}$, at least one of which is non-zero. You can assume $a_{ij} = a_{ji}$ by replacing both of a_{ij}, a_{ji} by $\frac{1}{2}(a_{ij} + a_{ji})$. (In this section we tacitly assume that the ground field $\mathbb{K}$ has characteristic $\neq 2$, so this replacement is possible.) Thus Q is completely described by the 3×3 symmetric matrix $A = (a_{ij})$ of coefficients; explicitly, setting $x = (x_1, x_2, x_3)$ we can write $Q = xAx^T$, where x^T denotes the transpose of the row vector x. The *rank* of Q is defined to be the rank of A. The rank is well defined; a projective mapping applied to Q will give rise to a matrix P^TAP (where P is a non-singular matrix) having the same rank as A.

The key to listing projective conics is the Lagrange method of 'completing the square' we met when discussing affine conics. The basic case is when at least one of x^2, y^2, z^2 appears in $Q(x, y, z)$. Select a variable (x say) whose square appears, think of $Q(x, y, z)$ as a quadratic in the single variable x (with coefficients involving y, z) and then 'complete the square' to obtain the sum of a square and a quadratic form in y, z alone; you then repeat the process to the latter quadratic form to express it as a sum of squares. Using this process we establish

Lemma 11.4 *Any conic Q in $P\mathbb{C}^2$ is projectively equivalent to one of the three forms in Table 11.1.*

Note that the rank determines the type of the conic. Of course these three types are also distinguished on the basis of factorization, since the irreducible type has a single component of multiplicity 1; the line-pair

Table 11.2. *The five types of real projective conic*

normal form	type	rank
$x^2 + y^2 + z^2$	empty irreducible conic	3
$x^2 + y^2 - z^2$	non-empty irreducible conic	3
$x^2 + y^2$	imaginary line-pair	2
$x^2 - y^2$	real line-pair	2
x^2	repeated line	1

has two distinct components, each of multiplicity 1; and the repeated line has a single component of multiplicity 2.

Lemma 11.5 *Any conic Q in $\mathbb{PR}^2$ is projectively equivalent to one of the five forms in Table 11.2.*

As in the complex case the rank distinguishes the two irreducible types from the two line-pairs, and from the repeated lines. The two irreducible types are distinguished from each other by their zero sets, as are the two line-pairs.

Example 11.3 Consider the conic $Q = x^2 + 5y^2 + 8z^2 + 2xy + 2xz - 6yz$ in $\mathbb{PR}^2$. You can select any of x, y, z; we will take x. Write Q as a quadratic in x, namely $Q = x^2 + 2x(y+z) + (5y^2 - 6yz + 8z^2)$. Completing the square, we get $Q = (x+y+z)^2 + (4y^2 - 8yz + 7z^2)$. You now work on the quadratic form $4y^2 - 8yz + 7z^2$: completing the square we get $4(y-z)^2 + 3z^2$. Thus $Q = (x+y+z)^2 + 4(y-z)^2 + 3z^2$. In other words $Q = X^2 + Y^2 + Z^2$ (so is of empty irreducible type) under the change of coordinates

$$X = x + y + z, \quad Y = 2(y - z), \quad Z = \sqrt{3}z.$$

What happens if none of x^2, y^2, z^2 appear in Q, so there are only cross terms xy, yz, zx? We use exactly the same trick adopted in the affine case. Suppose xy appears. The trick is to replace x, y by $x+y$, $x-y$ in Q (a projective change of coordinates) to force x^2, y^2 to appear in the formula, and then proceed as above.

Example 11.4 Consider the conic $Q = xy + yz + zx$ in $\mathbb{PR}^2$. Replacing x, y by $x+y$, $x-y$ we get $Q = x^2 - y^2 + 2xz = (x+z)^2 - y^2 - z^2$. Thus $Q = X^2 - Y^2 - Z^2$, where $X = x+z$, $Y = y$, $Z = z$ is a projective change of coordinates, so is of non-empty irreducible type.

Exercises

11.3.1 Classify the following conics in $\mathbb{PR}^2$, up to projective equivalence.

(i) $x^2 + 2z^2 + 2yz + 2xy$
(ii) $x^2 + y^2 + z^2 - yz + xz + xy$
(iii) $x^2 + y^2 + 4z^2 - 4yz - 4xz + 2xy$
(iv) $x^2 + 2y^2 + 9z^2 - 2yz + 2xz + 2xy$.

11.3.2 Classify the following conics in $\mathbb{R}^2$ up to affine equivalence. Which, if any, of these are affinely equivalent when regarded as conics in $\mathbb{C}^2$? Which, if any, of the associated projective conics are projectively equivalent in $\mathbb{PR}^2$? And which, if any, of the associated projective conics are projectively equivalent in $\mathbb{PC}^2$?

(i) $2x^2 + 21y^2 - 12xy - 4x + 18y$
(ii) $4x^2 + 12xy + 9y^2 - 28x - 48y + 48$
(iii) $4y^2 - 12xy - 48x + 20y + 17$.

11.4 Affine and Projective Equivalence

We finish this chapter by relating more closely the concepts of affine and projective equivalence. The relation between affine maps of $\mathbb{K}^2$ and projective maps of $\mathbb{PK}^2$ is easily described. For the sake of definiteness, let us take the line at infinity to be $z = 0$, so we can identify $\mathbb{K}^2$ with the set of points $(x : y : 1)$ in $\mathbb{PK}^2$. With this identification, an affine mapping has the form $\phi(x, y, 1) = (X, Y, 1)$ where $X = px+qy+\alpha$, $Y = rx+sy+\beta$, and $ps - qr \neq 0$.

Lemma 11.6 *Any affine map ϕ of $\mathbb{K}^2$ extends to a projective map Φ of $\mathbb{PK}^2$ mapping $z = 0$ to itself, and conversely, the restriction to $\mathbb{K}^2$ of any projective map Φ of $\mathbb{PK}^2$ mapping $z = 0$ to itself is an affine map ϕ. Further, let f, g be affine equivalent curves in $\mathbb{K}^2$ via an affine mapping ϕ, then the corresponding projective curves F, G in $\mathbb{PK}^2$ are projectively equivalent via Φ.*

Proof The formula $\Phi(x, y, z) = z\phi(x/z, y/z)$ clearly defines a projective map of $\mathbb{PK}^2$ taking $z = 0$ to itself. Conversely, any projective map of $\mathbb{PK}^2$ arises from an invertible linear map Φ of $\mathbb{K}^3$ given by a formula $\Phi(x, y, z) = (X, Y, Z)$ where X, Y, Z are independent linear forms in x, y, z. The map Φ takes $z = 0$ to itself if and only if $Z = kz$ for some scalar $k \neq 0$. The restriction ϕ to $\mathbb{K}^2$ is obtained by setting $z = 1$, and yields an affine mapping of $\mathbb{K}^2$. The hypothesis for the second statement

Table 11.3. *Complex affine conics viewed projectively*

normal form	affine type	projective type
$y^2 = x$	parabola	irreducible
$y^2 = -x^2 + 1$	irreducible conic	irreducible
$y^2 = x^2$	line-pair	line-pair
$y^2 = 1$	parallel lines	line-pair
$y^2 = 0$	repeated line	repeated line

Table 11.4. *Real affine conics viewed projectively*

normal form	affine type	projective type
$y^2 = x$	parabola	non-empty irreducible
$y^2 = -x^2 + 1$	real ellipse	non-empty irreducible
$y^2 = -x^2 - 1$	imaginary ellipse	empty irreducible
$y^2 = x^2 + 1$	hyperbola	non-empty irreducible
$y^2 = x^2$	real line-pair	real line-pair
$y^2 = -x^2$	imaginary line-pair	imaginary line-pair
$y^2 = 1$	real parallel lines	real line-pair
$y^2 = -1$	imaginary parallel lines	imaginary line-pair
$y^2 = 0$	repeated line	repeated line

is that $g(x, y) = \lambda f(\phi(x, y))$ for some scalar $\lambda \neq 0$. Suppose that f, g are of degree d. Then F, G are projectively equivalent via Φ, since

$$G(x, y, z) = z^d g(x/z, y/z) = z^d \lambda f(\phi(x/z, y/z)) = \lambda F(\Phi(x, y, z)).$$

□

Thus we can paraphrase Lemma 11.6 by saying that affinely equivalent affine curves give rise to projectively equivalent projective curves. The simplest illustration is that affinely equivalent conics give rise to projectively equivalent projective conics. Since we have listed both affine and projective types of conics, it is interesting to marry the two listings. In the complex case, it is easily verified that the affine types give rise to the projective types listed in Table 11.3, and in the real case by Table 11.4.

Another consequence of Lemma 11.6 is that if f, g are affine equivalent curves in $\mathbb{K}^2$, then the corresponding projective curves F, G in $P\mathbb{K}^2$ will be projectively equivalent, via a projective equivalence that preserves the line at infinity, and hence F, G will have the same intersection patterns with that line. Thus, given an affine curve in $\mathbb{K}^2$, the intersection pattern of the corresponding projective curve in $P\mathbb{K}^2$ with the line at infinity provides an affine invariant. This provides one way of distinguishing affine curves of the same degree. It was exactly this approach we applied in Section 10.4 to see how affine conics could be distinguished by the way in which the corresponding projective curve met the line at infinity. A more subtle application, to irreducible affine cubics having a cusp, is provided by the following example. The point of the example is that there are *several* affine types of such curves; by contrast, we will see in Chapter 15 that in the projective case there is a *unique* type.

Example 11.5 Consider an irreducible affine cubic f in $\mathbb{R}^2$ having just one singular point, namely a cusp. By applying an affine map, we can assume that the cusp is at $O = (0,0)$, and that the cuspidal tangent is $y = 0$. Thus our cubic has the form $y^2 = C(x, y)$ in $\mathbb{R}^2$ where $C(x, y)$ is a binary cubic form. The associated projective curve $y^2z = C(x, y)$ meets the line at infinity $z = 0$ at the points $(x : y : 0)$ for which $(x : y)$ is a root of the equation $C(x, y) = 0$. Note that y is not a factor of $C(x, y)$, else the curve reduces, thus the roots of $C(x, y) = 0$ all have the form $(x : 1)$ for some x, and the intersection pattern depends on the roots of the cubic $C(x, 1)$ in one variable. The general situation is that there are three complex roots, at least one of which is real; either there is just one real root, or there are three distinct real roots. Exceptionally, we can have two real roots, one of multiplicity two, or just one real root of multiplicity three. The geometric distinctions are illustrated in Figure 11.1. For instance, the semicubical parabola $y^2 = x^3$ corresponds to $C = x^3$ with one real root of multiplicity 3, whilst the cissoid of Diocles $y^2 = x(x^2 + y^2)$ corresponds to $C = x(x^2 + y^2)$ with one real root of multiplicity 1; thus these affine cubics cannot be affine equivalent.

This example represents a first step towards listing real affine cubics into distinct 'types', a study undertaken by Isaak Newton nearly three centuries ago, and representing one of the first serious excursions into curve theory. Newton's work was subsequently developed by other mathematicians, and is still an object of interesting research. Finally, here is yet another consequence of Lemma 11.6 which underpins many of the later assertions in this book.

Lemma 11.7 *Projective intersection numbers are invariant in the following sense. Suppose we have a projective map* Φ *of* $P\mathbb{K}^2$ *taking the point* P, *the*

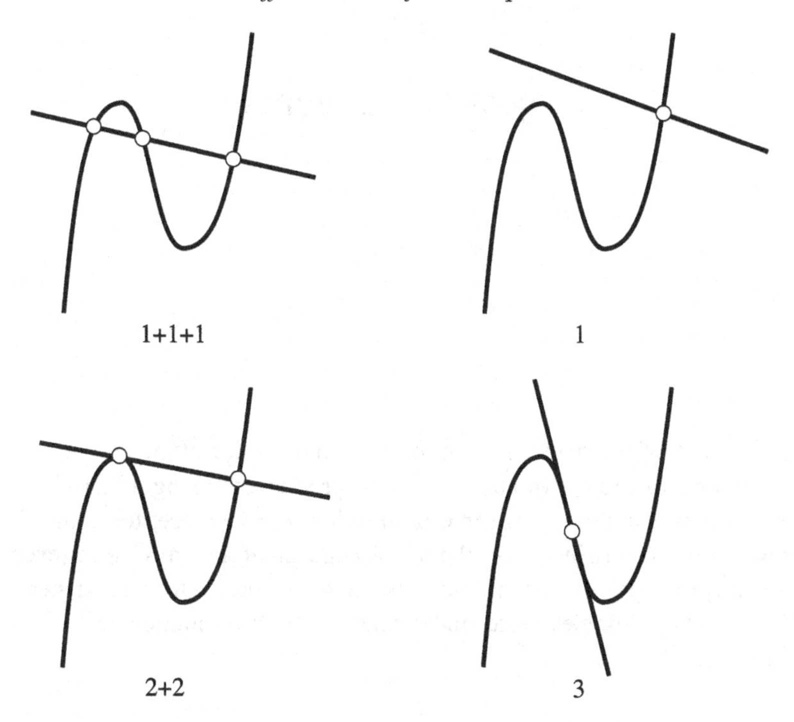

Fig. 11.1. Root patterns of cubics in one variable

curve F and the line L to the point P', the curve F' and the line L'. Then $I(P,F,L) = I(P',F',L')$.

Proof In the domain of Φ dehomogenize with respect to one of the variables x, y, z to obtain a point p, a curve f, and a line l in the affine plane $\mathbb{K}^2$. Recall from Lemma 10.5 that $I(P,F,L) = I(p,f,l)$. Likewise, in the target of Φ dehomogenize to obtain a point p', a curve f', and a line l' in the affine plane $\mathbb{K}^2$ for which $I(P',F',L') = I(p',f',l')$. Thus it remains to show that $I(p,f,l) = I(p',f',l')$, which follows immediately from Lemma 11.6 and Lemma 6.3. □

12
Projective Tangents

The concept of 'tangent', developed in Chapter 7 for affine curves, can be extended to projective curves, via the process of taking affine views. The result is a number of geometric insights. For instance, for conics it throws light (surprisingly) on the affine concept of a 'centre', and (even more surprisingly) on the metric concept of a 'focus'. And for general affine curves, it enables us to understand better the common feature of 'asymptotes'.

12.1 Tangents to Projective Curves

Tangent lines to projective curves are defined by analogy with the affine case. Let P be a point of multiplicity m on a projective curve F in $\mathbb{PK}^2$. Then $I(P,F,L) \geq m$ for every line L through P. We say that a line L is *tangent* to F at P, or a *tangent line* to F at P, when $I(P,F,L) > m$. The concept of tangency is invariant under projective maps in the following sense. Suppose that under a projective map P, F, L correspond to P', F', L', then L is tangent to F at P if and only if L' is tangent to F' at P'. That follows immediately from the fact (Lemma 11.7) that intersection numbers (of curves with lines) are invariant under projective maps. Generally, we say that two curves F, G are *tangent* at the point P when they have a common tangent at P. In view of the above remarks the general concept of tangency is invariant under projective changes of coordinates.

We can find tangents in practice via a lemma from the previous chapter. Suppose we dehomogenize with respect to one of the variables x, y, z to obtain a point p, a curve f, and a line l in the affine plane $\mathbb{K}^2$. Then by Lemma 10.5, $I(P,F,L) = I(p,f,l)$. Thus points p of multiplicity m on f correspnd to points P of multiplicity m on F, and tangents l to f at p

correspond to tangents L to F at P. In practice, we find the tangents to the affine curve f, and then 'projectivize' them. In particular, we see that at a point P of multiplicity m on F there are $\leq m$ tangents, which can be counted with multiplicities. Over the complex field, there are exactly m tangents, counted properly. In the next section, we will amplify these remarks in the important special case when P is a simple point.

Exercises

12.1.1 Let P be a point on a real curve F in $\mathbb{PC}^2$, and let L be a tangent to F at P. Show that the complex conjugate line $\overline{L}$ is a tangent to F at the complex conjugate point $\overline{P}$.

12.2 Tangents at Simple Points

As in the affine case, the simplest possible case of the above is the case $m = 1$ of a simple point P on a curve F. Here too there is an explicit formula for the tangent line, exhibiting greater symmetry than in the affine case.

Lemma 12.1 *Let P be a simple point on a curve F in $\mathbb{PK}^2$. The unique tangent to F at P has equation*

$$xF_x(P) + yF_y(P) + zF_z(P) = 0.$$

Proof It is no restriction to assume that $P = (a : b : 1)$, so we can dehomogenize with respect to z to obtain an affine curve $f(x, y) = F(x, y, 1)$. By Section 7.2, the tangent to f at $p = (a, b)$ has equation

$$(x - a)f_x(p) + (y - b)f_y(p) = 0$$

and hence the projective tangent is

$$(x - az)F_x(P) + (y - bz)F_y(P) = 0.$$

The Euler Lemma (with $x = a$, $y = b$, $z = 1$) tells us that

$$aF_x(P) + bF_y(P) + 1F_z(P) = 0.$$

Combining the last two displayed relations, we see that we can write the projective tangent in the desired form. □

Example 12.1 Let $P = (a : b : c)$ be a point on the line $F = \alpha x + \beta y + \gamma z$ in $\mathbb{PK}^2$. Then $F_x = \alpha$, $F_y = \beta$, $F_z = \gamma$ and the tangent at P to the line is defined by $xF_x(P) + yF_y(P) + zF_z(P) = 0$, i.e. $\alpha x + \beta y + \gamma z = 0$. Thus the tangent to a line at any point is the line itself.

Example 12.2 For conics, there is a particularly useful formula for the tangent line. Let $P = (\alpha : \beta : \gamma)$ be a simple point on the conic V in $\mathbb{PK}^2$. Then the tangent line to V at P has equation $\alpha V_x + \beta V_y + \gamma V_z = 0$. (Exercise 12.2.1.) For instance $P = (1 : 0 : 1)$ is simple on the projective conic $V = x^2 + y^2 - z^2$ in $\mathbb{PR}^2$, and the tangent at P is $x - z = 0$. In the affine view $z = 1$, that says that the tangent to the circle $x^2 + y^2 = 1$ at $(1, 0)$ is $x = 1$.

Exercises

12.2.1 Let $P = (\alpha : \beta : \gamma)$ be a simple point on a projective conic V in $\mathbb{PK}^2$. Show that the tangent to V at P is $\alpha V_x + \beta V_y + \gamma V_z = 0$.

12.3 Centres viewed Projectively

In this (and the next) section we return to a recurrent theme of this book, namely the geometry of conics. Our objective is to show that two geometric concepts (namely centres and foci), which are commonly discussed only in the affine situation, admit somewhat surprising projective interpretations. These provide excellent applications of the concept of tangency at a simple point. Let us start with the concept of 'centre'. Recall from Chapter 5 that the condition for an conic v in $\mathbb{R}^2$ to have a unique centre is that the invariant $\delta \neq 0$. However, we saw in Section 10.4 that this is precisely the condition for the corresponding projective conic V in $\mathbb{PC}^2$ to meet the line at infinity $z = 0$ in two distinct points. That prompts us to enquire whether there is a direct connexion between the centre of an affine conic v and the points at infinity on the corresponding projective conic V. We will assume v (and hence V) is irreducible: Lemma 10.9 tells us that every point on V is then simple, so the tangent is given by the formula in Lemma 12.1.

Lemma 12.2 *Let v be an irreducible conic in $\mathbb{R}^2$, and let V be the corresponding projective conic in $\mathbb{PC}^2$. Assume v has a unique centre $p = (\alpha, \beta)$. Then $P = (\alpha : \beta : 1)$ is the point of intersection of the tangents to V at its points at infinity.* (Figure 12.1.)

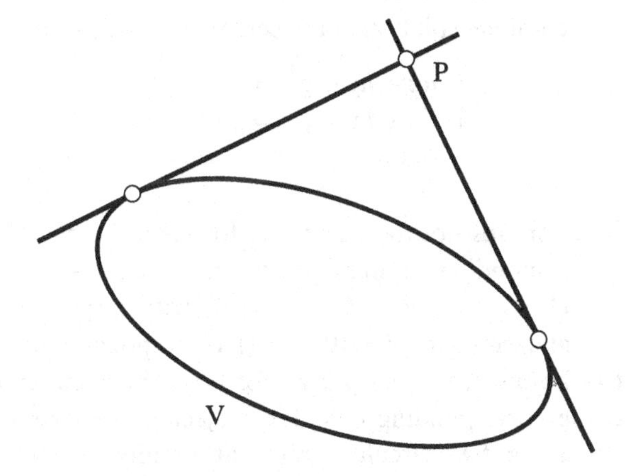

Fig. 12.1. Centres viewed projectively

Proof Set $v = ax^2+2hxy+by^2+2gx+2fy+c$, so the associated complex projective conic is given by the formula

$$V = ax^2 + 2hxy + by^2 + 2gzx + 2fyz + cz^2.$$

Recall from Lemma 10.10 that a conic in $P\mathbb{C}^2$ is irreducible if and only if it is non-singular. In particular, the irreducible conic V is non-singular. And recall also from Section 10.4 that (the projectivization of) a central conic meets the line at infinity $z = 0$ in two distinct points. The conic V meets $z = 0$ when $ax^2 + 2hxy + by^2 = 0$. Let $(x_1 : y_1)$, $(x_2 : y_2)$ be the distinct complex ratios satisfying this binary quadratic, so the points at infinity on V are $A = (x_1 : y_1 : 0)$, $B = (x_2 : y_2 : 0)$. These points are necessarily simple on V, and the tangents at A, B to V are

$$\begin{cases} x_1V_x + y_1V_y &= 0 \\ x_2V_x + y_2V_y &= 0. \end{cases}$$

Think of these as two linear equations in the unknowns V_x, V_y. The determinant of the 2×2 matrix of coefficients is $x_1y_2 - x_2y_1$, which is $\neq 0$ since A, B are distinct points in $P\mathbb{C}^2$. Thus the solution is given by

$$\begin{cases} 0 &= V_x &= ax + hy + gz \\ 0 &= V_y &= hx + by + fz. \end{cases}$$

Note that $z \neq 0$; indeed if $z = 0$, the equations only have the trivial solution $x = 0$, $y = 0$ since the determinant of the coefficients is $ab-h^2 = \delta \neq 0$. We can therefore assume that $z = 1$, yielding the following equations, whose unique solution is the centre $p = (\alpha, \beta)$ of v.

$$\begin{cases} ax + hy + g &= 0 \\ hx + by + f &= 0. \end{cases}$$

□

This result is perhaps not too surprising for (say) the hyperbola $x^2 - y^2 = 1$. The corresponding complex projective conic $x^2 - y^2 = z^2$ meets $z = 0$ at $A = (1 : 1 : 0)$, $B = (1 : -1 : 0)$ with respective tangents $y = x$, $y = -x$ intersecting at $P = (0 : 0 : 1)$, corresponding to the centre $p = (0, 0)$. It is, however, very surprising for (say) the circle $x^2 + y^2 = 1$. In that case, the corresponding complex projective conic $x^2 + y^2 = z^2$ meets $z = 0$ in the two circular points at infinity $I = (1 : i : 0)$, $J = (1 : -i : 0)$ with tangents $x + iy$, $x - iy$ intersecting at $P = (0 : 0 : 1)$, corresponding to the centre $p = (0, 0)$. This is a remarkable illustration of how the geometry of a real affine curve can be better understood via the processes of complexification and projectivization.

Exercises

12.3.1 For a *general* circle in $\mathbb{R}^2$, verify that the centre corresponds to the intersection of the tangents at the points $I = (1 : i : 0)$, $J = (1 : -i : 0)$ of the corresponding complex projective curve.

12.4 Foci viewed Projectively

In Example 1.6, we recalled the classical *metrical* definition of the 'focus' for a standard conic in $\mathbb{R}^2$. The concept can be extended to arbitrary curves f in $\mathbb{R}^2$ as follows. For any point P in $\mathbb{PC}^2$ distinct from the circular points at infinity I, J, we define the *isotropic lines* through P to be the lines joining P to I, J. Note that if P is a real point then the isotropic lines are complex conjugate. Now let $p = (a, b)$ be any point in $\mathbb{R}^2$, and let $P = (a : b : 1)$ be the corresponding point in $\mathbb{PC}^2$. (So P is real, and distinct from I, J.) Then p is a *focus* for a curve f in $\mathbb{R}^2$ when both the isotropic lines through P are tangent to the corresponding curve F in $\mathbb{PC}^2$.

Example 12.3 Consider a general circle f in $\mathbb{R}^2$. The condition for $p = (a, b)$ to be a focus is that the isotropic lines through $P = (a : b : 1)$ should be tangent to the corresponding complex projective curve F. Since a line meets an irreducible conic in just two points (counted properly) that means that the isotropic lines are the tangents to F at I, J. Thus P is the intersection of the tangents at I, J and by Exercise 12.3.1 corresponds to the *centre* p of f. Thus circles have a unique focus, namely their centre.

More generally, to find the foci of a real affine curve f we need to determine the tangents L to the corresponding complex projective curve F through I and determine their intersections with the complex conjugate tangents $\overline{L}$ to F through J. That raises the question of how many tangents there are to F passing through a given point in $P\mathbb{C}^2$. For conics, the answer is provided by the following lemma, of independent interest.

Lemma 12.3 *Let V be an irreducible conic in $P\mathbb{C}^2$, and let P be any point not lying on V. There are exactly two distinct lines through P tangent to V.*

Proof The proof is based on the result established in Section 11.3, namely that any irreducible conic in $P\mathbb{C}^2$ is projectively equivalent to the conic with normal form $F = x^2 + y^2 + z^2$. Since projective equivalences preserve tangencies, it suffices to establish the result for the normal form. The tangent line at $(X : Y : Z)$ to $x^2 + y^2 + z^2$ has equation $Xx + Yy + Zz = 0$, and the condition for this to pass through a given point $P = (\alpha : \beta : \gamma)$ is that $\alpha X + \beta Y + \gamma Z = 0$. One of α, β, γ is $\neq 0$. By symmetry, we can suppose $\gamma \neq 0$. Then $Z = -(\alpha X + \beta Y)/\gamma$, and substituting in $F = 0$ we get

$$(\alpha^2 + \gamma^2)X^2 + 2\alpha\beta XY + (\beta^2 + \gamma^2)Y^2 = 0$$

a binary quadratic with ≤ 2 solutions for the ratio $(X : Y)$. This only has a repeated root when the discriminant vanishes, giving

$$0 = 4\alpha^2\beta^2 - 4(\alpha^2 + \gamma^2)(\beta^2 + \gamma^2) = 4\gamma^2(\alpha^2 + \beta^2 + \gamma^2)$$

Since $\gamma \neq 0$ that reduces to $\alpha^2 + \beta^2 + \gamma^2 = 0$; but that is impossible as we assumed P does not lie on F. The result follows. □

Consider now a parabola v in $\mathbb{R}^2$, so the corresponding complex projective conic V in $P\mathbb{C}^2$ is tangent to the line at infinity $z = 0$ at some real point K. By Lemma 12.3 there are exactly two lines tangent to V through I, and exactly two lines tangent to V through J. In each case, $z = 0$ is one of the lines, so there is a line L distinct from $z = 0$ through I tangent to V, and its complex conjugate $\overline{L}$ is a line distinct from $z = 0$

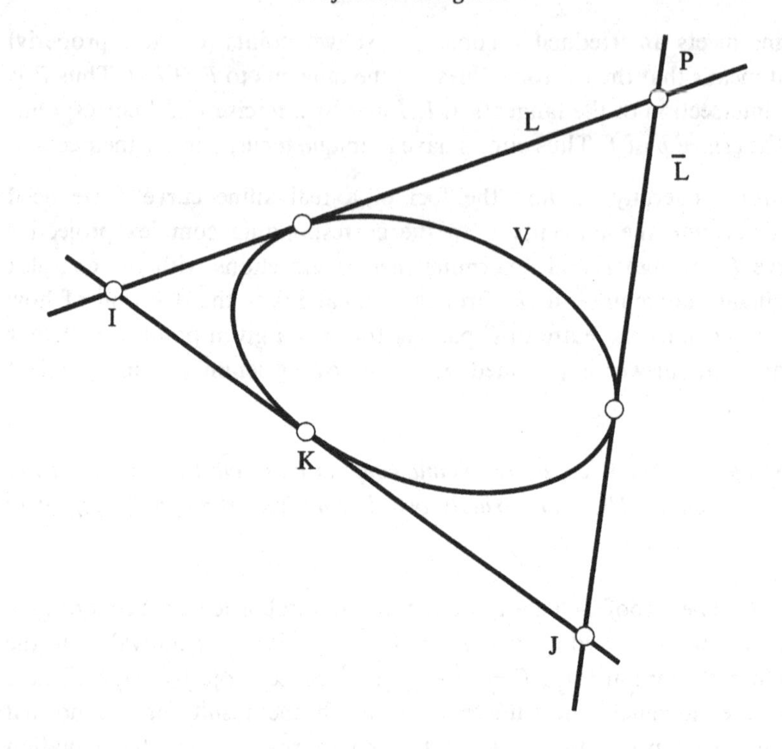

Fig. 12.2. Focus of a parabola

through J tangent to V. (Figure 12.2.) The intersection P of L, $\overline{L}$ is then the unique focus for v. The line in $\mathbb{R}^2$ corresponding to the line in $\mathbb{PC}^2$ joining P to K is the *axis* of the parabola v.

Example 12.4 We will determine the focus of the standard parabola $v = y^2 - 4ax$ with $a > 0$. The corresponding projective conic $V = y^2 - 4axz$ passes through neither I nor J, and is tangent to $z = 0$ at $K = (1 : 0 : 0)$. The tangent at $(X : Y : Z)$ is $2aZx - Yy + 2aXz = 0$, and passes through I if and only if $Y = -2aiZ$. We can suppose $Z \neq 0$. Substituting in $Y^2 = 4aXZ$ we obtain $X = -aZ$, yielding the point $(-a : -2ia : 1)$ on V with tangent $x + iy - az = 0$ through I. Conjugating, we obtain the point $(-a : 2ia : 1)$ on V with tangent $x - iy - az = 0$ through J. The tangents intersect when $x - az = 0$, $iy = 0$, i.e. at $P = (a : 0 : 1)$. The focus is therefore the point $p = (a, 0)$, agreeing with the metrical definition. And the axis is the line in $\mathbb{R}^2$ corresponding to the line in $\mathbb{PC}^2$ joining P and K, i.e. the line $y = 0$.

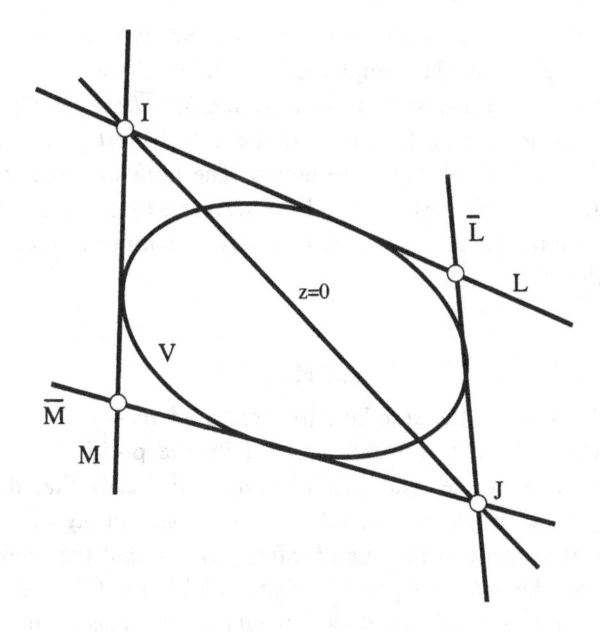

Fig. 12.3. Foci of ellipses and hyperbolas

Assume now that v is either an ellipse or a hyperbola (so V is not tangent to the line at infinity $z = 0$) but not a circle (so V does not pass through the circular points at infinity I, J). Then by Lemma 12.3 there are through each of I, J two distinct lines tangent to V. Indeed, if L, M are the lines through I tangent to V, then the complex conjugate lines $\overline{L}$, $\overline{M}$ are the lines through J tangent to V. There are therefore two foci, namely the points corresponding to the intersections of L, $\overline{L}$ and M, $\overline{M}$. (Figure 12.3.) The *axes* of v are defined to be the line in $\mathbb{R}^2$ joining the two foci, and the line in $\mathbb{R}^2$ corresponding to the line in $P\mathbb{C}^2$ joining the intersections of L, $\overline{M}$ and $\overline{L}$, M.

Example 12.5 We will determine the foci of the standard ellipse $v = b^2x^2 + a^2y^2 - a^2b^2$, where $0 < b < a$, with *eccentricity* the positive real number e defined by $a^2e^2 = a^2 - b^2$. The corresponding complex projective conic is $V = b^2x^2 + a^2y^2 - a^2b^2z^2$. The tangent to V at $(X : Y : Z)$ is $xb^2X + ya^2Y - a^2b^2zZ = 0$, and passes through I when $Y = ib^2X/a^2$. Substituting in the projective conic gives $Z = \pm Xe/a$. Thus the points $(X : Y : Z)$ on the conic where the tangents pass through I are $P = (a^2 : ib^2 : ae)$ and $Q = (a^2 : ib^2 : -ae)$, and the tangents at these points are $L = x + iy - ez$ and $M = x + iy + ez$. The tangents through J are the complex conjugate lines $\overline{L} = x - iy - ez$ and $\overline{M} = x - iy + ez$. The intersections of L, $\overline{L}$ and M, $\overline{M}$ are readily checked to be $F^{\pm} = (\pm ae : 0 : 1)$. It follows that the foci are $(\pm ae, 0)$, and one axis is the line $y = 0$ joining these points. The points of intersection of L, $\overline{M}$ and $\overline{L}$, M are $(0 : \pm ie : 1)$, and the second axis corresponds to the line $x = 0$ joining them. Again, the projective definitions agree with the metrical definitions.

Exercises

12.4.1 Take h to be the standard hyperbola $x^2/a^2 - y^2/b^2 = 1$ in $\mathbb{R}^2$, where $a > 0$, $b > 0$ with *eccentricity* the positive real number e defined by the relation $a^2e^2 = a^2 + b^2$. Verify that the corresponding projective conic H does not pass through I or J, and is not tangent to the line at infinity. Show that the points on H where the tangents pass through I and J are $(a^2 : -ib^2 : \pm ae)$, and that the corresponding tangents are the lines $x + iy \pm aez = 0$ and $x - iy \pm aez = 0$. Deduce that the foci of h are the points $(\pm ae, 0)$.

12.5 Tangents at Singular Points

Here are some examples of finding tangents to projective curves at singular points.

Example 12.6 The projective curve $F = y^2z^2 - x^2(z^2 - x^2)$ in $\mathbb{PR}^2$ has two singular points, at $P = (0 : 0 : 1)$ and $Q = (0 : 1 : 0)$. At P, dehomogenize with respect to z to obtain the affine curve $f = y^2 - x^2 + x^4$ with LOT $y^2 - x^2$, hence tangents $y = \pm x$. At Q, dehomogenize with respect to y to obtain the affine curve $g = z^2 - x^2(z^2 - x^2)$ with LOT $2z^2$, hence a repeated tangent $z = 0$.

Example 12.7 It is easily checked that the quartic $F = (x^2 - y^2)^2 + (2x^2 - 6y^2)z^2$ in $\mathbb{PC}^2$ has exactly three singular points $P = (1 : 1 : 0)$, $Q = (1 : -1 : 0)$, $R = (0 : 0 : 1)$, all of multiplicity 2. We will find the tangents to F at P. Dehomogenize by setting $x = 1$ to obtain an affine curve $f = (1 - y^2)^2 + (2 - 6y^2)z^2$ in the (y, z)-plane with singular point $p = (1, 0)$. Now translate the singular point to the origin by setting $y = Y + 1$, $z = Z$ to obtain an affine curve $g = 4(Y^2 - Z^2) + \text{HOT}$ in the

(Y, Z)-plane with singular point at $(0,0)$. The tangents to g at the origin are $Y = \pm Z$, and hence the tangents to f at p are $1 - y \pm z = 0$. Finally, homogenizing with respect to x we obtain the tangents $x - y \pm z = 0$ to F at P.

The various distinctions made for singular points in the affine case carry through to the projective case. Thus a point P of multiplicity m on F is *ordinary* when there are m distinct tangents to F at P. A double point which is ordinary is called a *node*; and a double point at which there is a repeated tangent is called a *cusp*. In the case of a node P, there are further distinctions to be made over the real field: the tangent lines are either both real (and P is a *crunode*) or complex conjugate (and P is an *acnode*).

Example 12.8 According to Example 10.7 the projective eight-curve has singular points at $P = (0 : 0 : 1)$ and $Q = (0 : 1 : 0)$: the computations show that P is a node (with tangents $y = \pm x$) and Q is a cusp (with cuspidal tangent $z = 0$).

Example 12.9 Consider the cubic curve $f = y(x^2 + y^2 - 2y)$ in $\mathbb{R}^2$ comprising the x-axis and the circle of radius 1 centred at the point $(0,1)$. The corresponding curve in $P\mathbb{R}^2$ is $F = x(x^2 + y^2 - 2yz)$ having exactly one singular point at $P = (0 : 0 : 1)$. Since $f = -2y^2 + \text{HOT}$, we see that P is a double point, with repeated tangent $y = 0$, hence a cusp.

Example 12.10 $y^2z = x^3 \pm x^2z$ has a unique double point $P = (0 : 0 : 1)$. For the $+$ sign, there are distinct real tangents $y = \pm x$, hence a crunode, and for the $-$ sign, there are distinct complex conjugate tangents $y = \pm ix$, hence an acnode. The affine views $z = 1$ of these curves were illustrated in Figure 7.3.

Exercises

12.5.1 Find the singular points, their multiplicities, and the tangents at the singular points, for the cubic in $P\mathbb{C}^2$ defined by $F = xz^2 - y^3 + xy^2$.

12.5.2 Find the singular points, their multiplicities, and the tangents at the singular points, for the cubic in $P\mathbb{C}^2$ defined by $F = (x + y + z)^3 - 27xyz$.

12.5.3 Find the singular points, their multiplicities, and the tangents at the singular points, for the quartic in $\mathbb{PC}^2$ defined by $F = xy^4 + yz^4 + xz^4$.

12.5.4 Find the singular points, their multiplicities, and the tangents at the singular points, for the quartic in $\mathbb{PC}^2$ defined by $F = x^2y^2 + y^2z^2 + z^2x^2$.

12.5.5 Find the singular points, their multiplicities, and the tangents at the singular points, for the quintic in $\mathbb{PC}^2$ defined by $F = (x^2 - y^2)^2 + (2x^2 - 6y^2)z^2$.

12.5.6 Find the singular points of the quartic curve in $\mathbb{PC}^2$ given by $F = (x^2 - z^2)^2 - y^2z(2y + 3z)$ and show that they are all nodes.

12.5.7 Find the singular points, their multiplicities, and the equations of the tangents at the singular points, for the quartic curve F in $\mathbb{PC}^2$ defined by $F = x^2y^2 + 36xz^3 + 24yz^3 + 108z^4$.

12.5.8 Show that the quartic curve $F = (x^2 + y^2 - zx)^2 - z^2(x^2 + y^2)$ in $\mathbb{PC}^2$ has exactly three singular points, all of which are simple cusps.

12.5.9 Let λ be a complex number and let F_λ be the quartic curve in $\mathbb{PC}^2$ given by $F_\lambda = x^4 + xy^3 + y^4 - \lambda zy^3 - 2x^2yz - xy^2z + y^2z^2$. Show that F_λ always has at least one singular point, a cusp. Further, show that there are exactly two further values of λ for which F_λ has in addition a node.

12.5.10 Show that the quintic $F = (x^2 - a^2z^2)^2y - (y^2 - b^2z^2)^2x$ in $\mathbb{PC}^2$ has exactly four singular points, all of which are nodes. (It is assumed that $a \neq 0, b \neq 0$.)

12.5.11 An irreducible quintic curve in the real projective plane $\mathbb{PR}^2$ is defined by $Q = (x^2 - z^2)^2y - (y^2 - z^2)^2x$. Verify that the quartic curve defined by the partial derivative Q_z reduces to an irreducible conic and two lines. Find the four real points of intersection of the conic with Q. Show that Q has no singular points on one line, exactly two singular points on the other line, and exactly two more singular points on the conic. Prove that all four singular points of Q are crunodes, and find the tangents at these points.

12.5.12 Show that the quintic curve in $\mathbb{PC}^2$ defined by $F = x^2y^3 + y^2z^3 + z^3x^2$ has exactly three singular points, namely an ordinary triple point A, a cusp B and a node C. Find the tangents to F at the singular points, and verify that none of them are components of F. Use your results to show that F cannot have a line component. (You may assume that two components of a curve in $\mathbb{PC}^2$ intersect in at least one point, and that any point of intersection is necessarily a singular point of the curve.)

12.5.13 Show that the sextic curve $F = (x^2 + y^2)^3 - 4x^2y^2z^2$ in $\mathbb{PC}^2$ has exactly three singular points, namely a real point A, and two complex conjugate points B and C. Show that A is a quadruple point with two repeated tangents. Show also that B, C are cusps, and that the line joining them is the common cuspidal tangent.

12.5.14 An irreducible quartic curve F in $\mathbb{PC}^2$ has exactly three singular points (all double points) at the vertices of the triangle of reference, so by Lemma 11.3 has the form

$$F = ay^2z^2 + bz^2x^2 + cx^2y^2 + 2xyz(fx + gy + hz)$$

where a, b, c are all non-zero. Show that the three double points are all cusps if and only if $f^2 = bc$, $g^2 = ca$ and $h^2 = ab$. In that case, show that F is projectively equivalent to the curve $y^2z^2+z^2x^2+x^2y^2-2xyz(x+y+z) = 0$, and verify that the three cuspidal tangents are concurrent.

12.5.15 Let a, b, c be non-zero *real* numbers. Show that the quartic curve $F = ax^2y^2+by^2z^2+cz^2x^2$ in $\mathbb{PC}^2$ has precisely three real singular points, all nodes. By appropriate scalings of the variables show that in $\mathbb{PC}^2$ the curve F is projectively equivalent to the curve $G = x^2y^2 + y^2z^2 + z^2x^2$. Why does this *argument* fail, when F is considered as a curve in $\mathbb{PR}^2$? Show that F, as a curve in $\mathbb{PR}^2$, has either one acnode and two crunodes, or three acnodes, and that both possibilities can occur. Deduce that F is not necessarily projectively equivalent in $\mathbb{PR}^2$ to the curve G.

12.6 Asymptotes

Let f be an affine curve in $\mathbb{R}^2$, and let F be the corresponding projective curve in $\mathbb{PR}^2$. An *asymptote* of f is defined to be a line in $\mathbb{R}^2$ whose projectivization in $\mathbb{PR}^2$ is a tangent to F at a point at infinity P. To find the asymptotes of f, first determine the points P where the associated projective curve F meets the line at infinity, then find the tangents to F at these points, and finally discard any tangent $z = 0$.

Example 12.11 For the standard rectangular hyperbola $f = x^2 - y^2 - 1$ in $\mathbb{R}^2$ the corresponding projective curve is $F = x^2 - y^2 - z^2$, which meets $z = 0$ at the points $P = (1 : 1 : 0)$ and $Q = (1 : -1 : 0)$. Both P, Q are

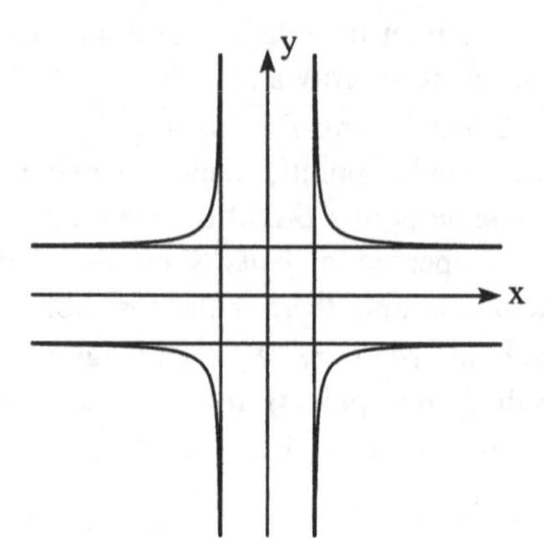

Fig. 12.4. Asymptotes of the cross curve

simple on F, with tangents $x + y = 0$, $x - y = 0$; thus we recover the familiar 'asymptotes' of elementary geometry.

Example 12.12 For the *cross* curve $f = x^2 + y^2 - x^2y^2$ in $\mathbb{R}^2$ the corresponding projective curve $F = x^2z^2 + y^2z^2 - x^2y^2$ meets $z = 0$ at $A = (1 : 0 : 0)$ (the point at infinity on the x-axis) and $B = (0 : 1 : 0)$ (the point at infinity on the y-axis). For the tangents at A, set $x = 1$ in F to get the affine curve $z^2 + y^2z^2 - y^2$, having a double point at $(0, 0)$ with distinct real tangents $y = \pm z$. (So A is a crunode.) Affinely the tangents at A are the asymptotes $y = \pm 1$. And for the tangents at B, set $y = 1$ in F to get the affine curve $x^2z^2 + z^2 - x^2$; this has a double point at $(0, 0)$ with distinct real tangents $x = \pm z$. (So B is also a crunode.) Affinely, the tangents at B are the asymptotes $x = \pm 1$. There are therefore four asymptotes $x = \pm 1$, $y = \pm 1$, illustrated in Figure 12.4. Incidentally, it is easily checked that F has only one further singular point, namely an acnode at $C = (0 : 0 : 1)$, giving rise to an isolated point at the origin on f. The zero set of f is easily sketched by writing the equation in the form $y^2 = x^2/(x^2 - 1)$.

Example 12.13 The projective curve corresponding to the affine curve $y = x^3$ in $\mathbb{R}^2$ is $yz^2 = x^3$, and meets the line at infinity at the unique point $P = (0 : 1 : 0)$. The point P is singular, indeed a cusp with cuspidal tangent $z = 0$. However, the tangent $z = 0$ is not the projectivization of a line in $\mathbb{R}^2$, so there are no asymptotes.

Exercises

12.6.1 In each of the following cases, find the asymptotes of the given curve f in $\mathbb{R}^2$.

(i) $f = x^3 - xy^2 - y$ (iv) $f = x^4 - x^2y^2 + y^2$
(ii) $f = x^2y^2 + x^2 - y^2$ (v) $f = (y - x^2)^2 - xy^3$
(iii) $f = x^4 + 4xy - y^4$ (vi) $f = x^5 + y^5 - 2xy^2$.

12.6.2 In each of the following cases, find the asymptotes of the given curve f in $\mathbb{R}^2$.

(i) $f = xy(x^2 - y^2) - (x^2 + y^2)$
(ii) $f = x^5 + y^5 - 2x^2 + 5xy + 2y^2$.

12.6.3 In each of the following cases, find the asymptotes of the given curve f in $\mathbb{R}^2$.

(i) $f = x^5 + y^5 - 5x^2y^2 + 3x - 3y$
(ii) $f = (2x + y)(x + y)^2 + (2x - y)(x + y) + (x - y)$
(iii) $f = x^4y + x^2y^3 + x^4 - y^3 + 1$.

12.6.4 Let $r(x) = p(x)/q(x)$ be a rational function with $p(x)$, $q(x)$ polynomials and $\deg p \leq \deg q$. Show that the asymptotes of $y = r(x)$ are the lines $x = c$, with c a root of $q(x) = 0$, and the line $y = d$ with $d = \lim_{x \to \infty} r(x)$.

13
Flexes

This chapter is devoted to the concept of 'flexes' on projective curves. There is a sense in which flexes are rather akin to singularities, and share their geometric significance; for instance, curves without line components have only finitely many flexes, and can provide insight into their geometry. Let P be a simple point on a curve F in $P\mathbb{K}^2$. Then the intersection number of the unique tangent line L to F at P with F is ≥ 2. In general, the intersection number $I(P,F,L)$ will be exactly two, but at exceptional points on the curve, one might expect that number to take higher values. That leads to the following definitions. The point P is a *flex* of F when $I(P,F,L) \geq 3$; that includes the case when $I(P,F,L) = \infty$. Further, P is an *ordinary* flex when $I(P,F,L) = 3$, and an *undulation* when $I(P,F,L) \geq 4$. Note that every point on a line is a flex, and that conics cannot possess flexes, so the concept is only of interest for curves of degree ≥ 3. Since intersection numbers are projective invariants, the concept of a flex (and the distinction between ordinary and higher flexes) is invariant under projective maps. Generally speaking, it is inefficient to go back to the definitions to find the flexes of a curve F.

Example 13.1 Consider the cubic curve $F = y^2z - x^3$ in $P\mathbb{C}^2$. It is easily verified that F has a unique singular point, at $Q = (0:0:1)$. By Lemma 12.1, the tangent L at any other point $P = (a:b:c)$ on F is $L = -3a^2x + 2bcy + b^2z$. We need to calculate $I(P,F,L)$. We can assume $b \neq 0$, for if $b = 0$ the relation $b^2c = a^3$ yields $a = 0$ as well, and we are at the singular point Q. Eliminating z between $L = 0$, $F = 0$ gives the binary cubic G defined by

$$bG = -b^3x^3 + 3a^2bxy^2 - 2b^2cy^3 = -(bx - ay)^2(bx + 2ay)$$

having $(a : b)$ as a zero of multiplicity ≥ 2, as it should. For P to be a flex we require $(a : b)$ to be a zero of multiplicity 3; that is the case if and only if the factors $bx - ay$, $bx + 2ay$ agree (up to scalar multiples), requiring $a = 0$, and hence $c = 0$ as well since $b^2c = a^3$. That yields $R = (0 : 1 : 0)$ as the unique flex on F.

The key to this chapter is a technique which reduces the question of finding flexes to that of finding the intersections of the curve with its associated 'Hessian curve'. That is the subject of the next section.

Exercises

13.0.1 Let F be an algebraic curve in $\mathbb{PC}^2$, and let P be an ordinary point on F, of multiplicity m. A tangent line L at P is *flexional* when $I(P, F, L) \geq m + 2$. (When $m = 1$, this means P is a flex.) Show that $F = x^4 + y^4 + xyz^2$ has just one singular point, an ordinary double point, and that both tangents are flexional.

13.1 Hessian Curves

The motivation is as follows. Let F be a curve in $\mathbb{PK}^2$. Write x_1, x_2, x_3 for homogeneous coordinates, and set $x = (x_1, x_2, x_3)$. The Maclaurin expansion of the polynomial F at P is

$$F(x) = F(P) + \sum F_i(P)x_i + \sum F_{ij}(P)x_ix_j + \cdots.$$

Let P be a simple point on F. Then $F(P) = 0$, and the leading term is $\sum F_i(P)x_i$, the unique tangent to F at P. Higher order information is provided by the quadratic terms. To any point P on F, we can associate the quadratic part of the Taylor series, namely the *Hessian* quadratic form

$$xF_2(P)x^T = \sum F_{ij}(P)x_ix_j$$

where the superscript T denotes transposition, and $F_2(P) = (F_{ij}(P))$ is the 3×3 symmetric *Hessian matrix* of second order partial derivatives, evaluated at P. Note that this form is identically zero if and only if all the coefficients $F_{ij}(P)$ vanish, i.e. if and only if P is a point of multiplicity ≥ 3 on F. The Hessian matrix is singular if and only if $H_F(P) = 0$, where H_F is the *Hessian determinant* $H_F(P) = \det F_2(P)$. The first thing to establish is that the vanishing of the Hessian determinant is invariant under projective mappings.

Lemma 13.1 *Suppose that under a projective map F becomes G, and P is mapped to Q, then $H_F(P) = 0$ if and only if $H_G(Q) = 0$.*

Proof A projective mapping is induced by an invertible linear map, $x \mapsto xA$ where A is an invertible 3×3 matrix. Thus the Hessian form $(F_{ij}(P))$ is replaced by $(AF_{ij}(P)A^T)$, and

$$\begin{aligned} H_G(Q) &= \det(G_{ij}(Q)) = \det(AF_{ij}(P)A^T) \\ &= \det A \det(F_{ij}(P)) \det A^T = (\det A)^2 \det(F_{ij}(P)) \\ &= (\det A)^2 H_F(P). \end{aligned}$$

The result is immediate as $\det A \neq 0$. □

Lemma 13.2 *Assume P is simple on F, then P is a flex if and only if $H_F(P) = 0$.*

Proof The concept of a flex is invariant under projective mappings: if under a projective map F becomes G, and P is mapped to Q, then P is a flex on F if and only if Q is a flex on G. By the previous lemma, we can therefore assume (by applying a suitably chosen projective map) that $P = (0 : 0 : 1)$, and that the tangent line L to F at P is $y = 0$. Thus, in the affine view $z = 1$, we have an affine curve

$$f(x, y) = \lambda y + (ax^2 + 2bxy + cy^2) + \text{HOT}$$

where $\lambda \neq 0$, and F is given by a formula of the form

$$F(x, y, z) = \lambda yz^{d-1} + (ax^2 + 2bxy + cy^2)z^{d-2} + \cdots.$$

The intersection of F with the tangent $y = 0$ at P corresponds to the root $(0 : 1)$ of the equation

$$F(x, 0, z) = ax^2z^{d-2} + \cdots = 0.$$

The condition for P to be a flex is that $I(P, F, L) \geq 3$, i.e. that the root $(0 : 1)$ should have multiplicity ≥ 3, i.e. that $a = 0$. It remains to show that this is equivalent to the condition $H_F(P) = 0$. That follows immediately from the following relation, which the reader is left to check.

$$H_F(P) = \begin{vmatrix} 2a & 2b & 0 \\ 2b & 2c & \lambda(d-1) \\ 0 & \lambda(d-1) & 0 \end{vmatrix} = -2a\lambda^2(d-1)^2.$$

□

Provided H_F is not identically zero, it is a form defining a projective curve of degree $3(d-2)$, the *Hessian* curve H_F associated to F. Thus the flexes of F are the intersections with H_F, simple on F. In practice, one finds the intersections of F, H_F and then eliminates those which are singular on F. We will illustrate the technique by re-doing Example 13.1.

Example 13.2 Consider again the cubic curve $F = y^2z - x^3$ in $P\mathbb{C}^2$ with unique singular point $Q = (0:0:1)$. The Hessian curve is

$$H_F = \begin{vmatrix} -6x & 0 & 0 \\ 0 & 2z & 2y \\ 0 & 2y & 0 \end{vmatrix} = 24xy^2.$$

Thus H_F vanishes when $xy^2 = 0$, i.e. on the lines $x = 0$, $y = 0$. The cubic F intersects $x = 0$ when $y^2z = 0$, i.e. when $y = 0$ or when $z = 0$. That yields the simple point $P = (0:1:0)$ and the singular point $Q = (0:0:1)$. The point P is thus the only intersection of F, H_F simple on F, hence the only flex of F, as we found in Example 13.1 by direct calculation.

Example 13.3 The cubic curve $F = x^3 + y^3 - xyz$ in $P\mathbb{C}^2$ is irreducible, and is easily checked to have a unique singular point at $Q = (0:0:1)$, namely a node with tangents $x = 0$, $y = 0$. We will determine its flexes. The Hessian curve is

$$H_F = \begin{vmatrix} 6x & -z & -y \\ -z & 6y & -x \\ -y & -x & 0 \end{vmatrix} = -2\{3(x^3 + y^3) + xyz\}.$$

The intersections of F, H_F are given by $x^3 + y^3 = 0$, $xyz = 0$. The latter relation yields $x = 0$, $y = 0$ or $z = 0$. Clearly, $x = 0$, $y = 0$ produce only the singular point Q. However $z = 0$ produces the three collinear points $A = (1:-1:0)$, $B = (1:-\omega:0)$, $C = (1:-\omega^2:0)$ where ω is a complex cube root of unity. The points A, B, C are simple on F, since they are distinct from Q, and hence are the flexes.

In Chapter 15 we will see that any two irreducible cubics in $P\mathbb{C}^2$ with just one node are projectively equivalent (the 'nodal' type) so that any such cubic has exactly three collinear flexes. Moreover, in Chapter 17 we will see that it is no coincidence that the flexes are collinear, there are underlying algebraic reasons why that has to be the case. In the next section we will look at a significantly more complex example, of considerable geometric interest.

Example 13.4 We frequently meet the case when f is a curve in $\mathbb{C}^2$ of degree $d \geq 3$, and F is the corresponding curve in $P\mathbb{C}^2$. We claim that the 'affine' Hessian h_f (meaning the equation of the Hessian H_F in the affine view $z = 1$) is given by

$$h_f = \begin{vmatrix} f_{xx} & f_{xy} & f_x \\ f_{yx} & f_{yy} & f_y \\ f_x & f_y & \frac{d}{d-1} f \end{vmatrix} = 0.$$

This can be seen as follows. In the determinantal expression for H_F, perform the column operations z.col $3 + y$.col $2 + x$.col 1, the row operations z.row $3 + y$.row $2 + x$.row 1, and simplify using the Euler Lemma. The claim follows on setting $z = 1$.

In principle, we expect to have at most finitely many flexes on a projective curve. Lines are an obvious exception. The next result (one of the very few in this text using the Implicit Function Theorem) shows that in a sense lines are the only exception.

Lemma 13.3 *Let F be an irreducible curve in $P\mathbb{C}^2$. Then the Hessian vanishes at every point of F if and only if F is a line.*

Proof When F is a line the Hessian vanishes identically, so certainly at every point of F. Conversely, suppose the Hessian vanishes at every point of F. In the next chapter, we will show that any irreducible curve F in $P\mathbb{C}^2$ has at most finitely many singular points. Since the zero set of F is infinite that means that F has a simple point P, at which the Hessian vanishes. By applying a projective mapping, we can assume that $P = (0:0:1)$, $F_y(0:0:1) \neq 0$. Set $f(x, y) = F(x, y, z)$, so $f(0,0) = 0$ and $f_y(0,0) \neq 0$. By the Implicit Function Theorem, there is an analytic function $y(x)$, defined on a neighbourhood of the origin in $\mathbb{C}$, with $y(0) = 0$ having the property that $f(x, y) = 0$ if and only if $y = y(x)$. Differentiating the identity $f(x, y(x)) \equiv 0$ with respect to x (via the Chain Rule) we obtain $y' = -f_x/f_y$, where the dash denotes differentiation with respect to x. Differentiating this relation with respect to x, we obtain $y'' = h_f/f_y^3$, after a few lines of computation, where h_f is the 'affine' Hessian. Since h_f vanishes identically on the zero set of f, we see that $y''(x) = 0$ for all x close to the origin in $\mathbb{C}$. It follows that $y(x) = ax + b$ for some complex numbers a, b. The condition $y(0) = 0$ means that $b = 0$, so $y(x) = ax$ for all x sufficiently close to the origin in $\mathbb{C}$. That means that the curve f meets the line $y = ax$ in infinitely many points, so the line is a component of f. But f is assumed irreducible, so f must be that line. □

Example 13.5 The proof of the above lemma shows that if the equation of a real affine curve can be written in the form $y = p(x)$ then the flexes are given by the condition $p''(x) = 0$. In this way, we recover the result of Example 7.4. By way of explicit illustration, consider the cubic curves $y(x^2+1) = 1$ (an example of Agnesi's versiera) and $y(x^2+1) = x$ (an example of the Serpentine). These cannot be distinguished by their singularities; in the real affine plane they are non-singular, and in the real projective plane both have exactly one singular point, an acnode. However, they can be distinguished by their flexes. Both curves can be written in the form $y = p(x)$ for an appropriate rational function $p(x)$. For Agnesi's versiera $p(x) = 1/(x^2+1)$, and the condition $p''(x) = 0$ reduces to $3x^2 - 1 = 0$, yielding *two* flexes. Likewise, for the Serpentine $p(x) = x/(x^2+1)$, and the condition $p''(x) = 0$ reduces to $x(x^2-3) = 0$, yielding *three* flexes. Since flexes are invariant under affine mappings the curves cannot be affinely equivalent.

Exercises

13.1.1 Show that the cubic curve $F = x^3 + y^3 - xy(x+y+z)$ in $P\mathbb{C}^2$ has a unique singular point, namely a node. Show that F has exactly three collinear flexes.

13.1.2 Show that $P = (0:0:1)$ is the unique singular point of the cubic curve $F = z(x^2+y^2) - y(3x^2-y^2)$ in $P\mathbb{C}^2$. Verify that the associated Hessian curve is $H_F = z(x^2+y^2) + 3y(3x^2-y^2)$. Deduce that F has exactly three collinear flexes.

13.1.3 Let F be a projective folium of Descartes, i.e. a cubic curve in $P\mathbb{R}^2$ given by a formula of the form $F = x^3 + y^3 - 3axyz$ for some scalar $a \neq 0$. Show that the Hessian of F is another folium of Descartes.

13.1.4 A cubic curve in $\mathbb{R}^2$ is defined by $f = xy^2 - y^3 + x$. Show that the corresponding projective curve F in $P\mathbb{R}^2$ has exactly one singular point A, namely an acnode on the line at infinity. Find the second point B where F meets the line at infinity, and show that f has a unique asymptote $x = y$. Verify that the Hessian curve $H_F = -8(xy^2 - 3yz^2 + xz^2)$. Show that F has exactly three flexes C, D, E on the line $4x = 3y$. Solve the equation $f = 0$ for x in terms of y, and hence give a rough sketch of the curve f.

13.1.5 Show that the cubic curve $F = x^3 + x^2z + y^2z$ in $P\mathbb{C}^2$ has a unique singular point $P = (0:0:1)$. Verify that the associated

Hessian curve is $H_F = -8(3xy^2 + x^2z + y^2z)$. Deduce that F has three distinct collinear flexes.

13.1.6 Show that the cubic curve $F = yz(x+3y+3z)+(y^3+z^3)$ in $\mathbb{PC}^2$ has just one singular point, a node at $P = (1:0:0)$. Verify that the Hessian curve is $H_F = 2yz(x+3y+3z)-6(y^3+z^3)$. Deduce that F has exactly three flexes, all on the line $x+3y+3z=0$.

13.1.7 Show that the cubic curve $F = 3x^2y - 4x^3 - 3x^2z + yz^2 - z^3$ in $\mathbb{PC}^2$ has just one singular point, namely a node. Show that F has Hessian $H_F = -F + 4x(z^2 - x^2)$. Deduce that F has three flexes, lying on the line $x - y + z = 0$.

13.1.8 Show that the cubic curve $F = xyz + (y+z)^3$ in $\mathbb{PC}^2$ has exactly three flexes, all of which lie on the line $x + 3y + 3z = 0$.

13.1.9 Show that the quartic curve $F = x^4+y^4+z^4$ in $\mathbb{PC}^2$ has precisely twelve flexes, and write down their coordinates.

13.1.10 Verify that $A = (1:0:0)$, $B = (0:1:0)$, $C = (0:0:1)$ are the only singular points of the quartic curve $F = x^2y^2 + y^2z^2 + z^2x^2$ in $\mathbb{PC}^2$. Verify that the associated Hessian curve H_F is given by the formula below, and hence show that F has no flexes.

$$H_F = 19x^2y^2z^2 - 3(x^2 + y^2 + z^2)F(x, y, z).$$

13.1.11 Let ω be a primitive complex cube root of unity. Verify that the points $A = (1:1:1)$, $B = (\omega:\omega^2:1)$, $C = (\omega^2:\omega:1)$ are singular points of the cubic curve $F = x^3+y^3+z^3-3xyz$ in $\mathbb{PC}^2$. Deduce that F reduces to three lines, one real and two complex conjugate. Show that the cubic curve $G = x^2y+y^2z+z^2x$ in $\mathbb{PC}^2$ is non-singular, and has Hessian F. Deduce that the *real* flexes of G are its intersections with the line $x + y + z = 0$.

13.1.12 Show that the flexes of the family of quartic curves in $\mathbb{PC}^2$ defined by $F_\lambda = x^2yz - y^3(2x+y) - \lambda x^4$ all lie on the cubic curve $x^2z + 2y^3 = 0$.

13.1.13 Let μ be a real number. A cubic curve in $\mathbb{PC}^2$ is defined by $F_\mu = (x+y+z)^3 + 6\mu xyz$. It is assumed that μ is chosen so that F_μ is non-singular. Verify that the Hessian of F_μ is the curve

$$H_\mu = (x+y+z)(x^2+y^2+z^2-2xy-2yz-2zx) - 2\mu xyz.$$

Deduce that the flexes of F_μ are the intersections of F_μ with the lines $x+y+z=0$, $x+\omega y+\omega^2 z=0$ and $x+\omega^2 y+\omega z=0$, where ω is a primitive complex cube root of 1. Use this fact to write down the coordinates of the real flexes of F_μ.

13.2 Configurations of Flexes

In Example 10.8, we introduced an interesting family of cubic curves, namely the Steiner cubics $F = \mu(x^3 + y^3 + z^3) + 3\lambda xyz$. We showed that F_λ is singular if and only if $\mu = 0$ or $\lambda^3 = -1$, in which cases F reduces to a triangle. We will determine the flexes of the Steiner cubics.

Example 13.6 We assume F is non-singular, so can assume $\mu = 1$, $\lambda^3 \neq -1$. The Hessian matrix is

$$(F_{ij}) = \begin{pmatrix} 6x & 3\lambda z & 3\lambda y \\ 3\lambda z & 6y & 3\lambda x \\ 3\lambda y & 3\lambda x & 6z \end{pmatrix}.$$

Computation yields

$$H_F = \det(F_{ij}) = 27(-2\lambda^2(x^3 + y^3 + z^3) + 2(4 + \lambda^3)xyz).$$

To find the intersections of F, H_F we have to solve the equations

$$\begin{cases} 0 = (x^3 + y^3 + z^3) + 3\lambda xyz \\ 0 = -2\lambda^2(x^3 + y^3 + z^3) + 2(4 + \lambda^3)xyz. \end{cases}$$

Note that both F, H_F are linear combinations of $x^3 + y^3 + z^3$ and xyz. They are scalar multiples of each other if and only if the 2×2 matrix of coefficients has zero determinant, i.e. if and only if (by computation) $\lambda^3 = -1$. In these cases, both curves reduce to the same triangle. Otherwise F is non-singular, and the intersections of F, H_F are given by $x^3 + y^3 + z^3 = 0$, $xyz = 0$. Suppose for instance that $x = 0$, then $y^3 + z^3 = 0$ and taking $z = 1$ we have $y^3 = -1$ so $y = -1$, $y = -\omega$ or $y = -\omega^2$ where ω is a primitive complex cube root of 1. Proceeding similarly for the cases $y = 0$, $z = 0$ we obtain the following array of flexes.

$$\begin{array}{lll} P_{00} = (0 : -1 : 1) & P_{01} = (0 : -\omega : 1) & P_{02} = (0 : -\omega^2 : 1) \\ P_{10} = (1 : 0 : -1) & P_{11} = (1 : 0 : -\omega) & P_{12} = (1 : 0 : -\omega^2) \\ P_{20} = (-1 : 1 : 0) & P_{21} = (-\omega : 1 : 0) & P_{22} = (-\omega^2 : 1 : 0). \end{array}$$

Note that the nine flexes in this example do not depend on the value of λ. Only three of them (those in the first column) are real points in $\mathbb{PC}^2$, and only the first two of those correspond to points in the affine view $z = 1$, namely $(0, -1)$ and $(-1, 0)$. Figure 13.1 illustrates these for the curve F with $\lambda = 0$. Note that the affine curve is affinely equivalent (under central reflexion in the origin) to the Fermat cubic $x^3 + y^3 = 1$.

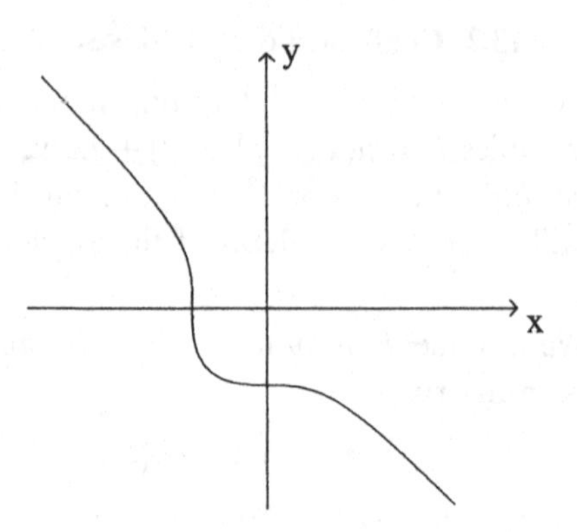

Fig. 13.1. Two flexes of the Fermat cubic

Lemma 13.4 *Any cubic curve G in $\mathbb{PC}^2$ which passes through the nine flexes P_{ij} of the Steiner cubics is itself a Steiner cubic.*

Proof Consider the conditions for G to pass through the points P_{00}, P_{01}, P_{02}, i.e. the points $(0 : -w : 1)$ with w a cube root of 1. Write a, b, c, d for the coefficients in G of the monomials y^3, y^2z, yz^2, z^3 not involving x. Then the conditions are that we have

$$\begin{cases} -a+b-c+d &= 0 \quad \text{(with } w=1\text{)} \\ -a+b\omega^2-c\omega+d &= 0 \quad \text{(with } w=\omega\text{)} \\ -a+b\omega-c\omega^2+d &= 0 \quad \text{(with } w=\omega^2\text{).} \end{cases}$$

A minor computation shows that $b = 0$, $c = 0$, $a = d$. Thus the terms in G which do not involve x must have the form $a(y^3 + z^3)$. Likewise the condition for G to pass through P_{10}, P_{11}, P_{12} is that the terms in G which do not involve y must have the form $a(z^3 + x^3)$. And the condition for G to pass through P_{20}, P_{21}, P_{22} is that the terms in G which do not involve z must have the form $a(x^3 + y^3)$. We conclude that G must have the form $G = \mu(x^3 + y^3 + z^3) + 3\lambda xyz$, for some λ, μ not both zero, and hence is a Steiner cubic. □

The configuration formed by the array of nine flexes in this example is particularly interesting. For instance, the points on the first, second, third rows lie on the lines $x = 0$, $y = 0$, $z = 0$; the points on the first, second, third columns lie on the lines $x + y + z = 0$, $\omega x + \omega^2 y + z = 0$, $\omega^2 x + \omega y + z = 0$; and the points on the two main diagonals lie on the lines $\omega x + y + z = 0$, $\omega x + \omega y + z = 0$. Indeed, the reader will easily check that *the line through any two of the flexes passes through exactly one more flex*. The nine flexes of the Steiner cubics therefore provide an example of what we will call a *nine point configuration*, namely a set of nine distinct points in $P\mathbb{K}^2$ with the property that the line through any two passes through exactly one more. Such a configuration is precisely that of the nine points in the affine plane over the field $\mathbb{Z}_3$, illustrated in Figure 2.2; indeed the notation has been chosen so that P_{ij} corresponds to the point (i, j) in Example 2.7. Generally, we can label the points of a nine point configuration as Q_{ij} with $0 \leq i, j \leq 2$ in such a way that the line through two points in the configuration corresponds to the line through the corresponding points in the affine plane over $\mathbb{Z}_3$.

It is clear that the concept of a nine point configuration is invariant under projective mappings. The next result tells us that all nine point configurations are projectively 'the same'.

Lemma 13.5 *For any nine point configuration Q_{ij} in $P\mathbb{C}^2$, there is a projective change of coordinates taking the points of the configuration to the 'standard' nine point configuration S_{ij} in $P\mathbb{C}^2$ given below, where $\alpha^2 = -3$.*

$$\begin{array}{lll} S_{00} = (-1 : 1 : 1) & S_{01} = (\alpha : 1 : 1) & S_{02} = (1 : 1 : 1) \\ S_{10} = (-1 : \alpha : 1) & S_{11} = (0 : 0 : 1) & S_{12} = (1 : -\alpha : 1) \\ S_{20} = (-1 : -1 : 1) & S_{21} = (-\alpha : -1 : 1) & S_{22} = (1 : -1 : 1). \end{array}$$

Proof As above, label the points in the configuration as Q_{ij} with $0 \leq i, j \leq 3$. The Four Point Lemma ensures that there is a (unique) projective map Φ taking $Q_{00}, Q_{02}, Q_{20}, Q_{22}$ respectively to the points $S_{00}, S_{02}, S_{20}, S_{22}$. (Clearly, both sets of four points are in general position.) We write S_{ij} for the image of Q_{ij} under Φ. What we have to show is that the remaining five points $S_{11}, S_{01}, S_{21}, S_{10}, S_{12}$ have the given homogeneous coordinates. Note first that S_{11} is the point of intersection of the line $y = x$ joining S_{20}, S_{02}, and the line $y = -x$ joining S_{00}, S_{22}, so that $S_{11} = (0 : 0 : 1)$. The point S_{01} lies on the line $y = z$ joining S_{00}, S_{02} so has the form $S_{01} = (\alpha : \beta : \beta)$ for some scalars α, β not both zero; likewise S_{21} lies on the line $y = -z$ joining S_{20}, S_{22} so has the form $S_{21} = (\gamma : -\delta : \delta)$ for some scalars γ, δ not both zero. However, S_{01}, S_{11}, S_{21} are collinear, which is easily checked to hold if and only if $\alpha\delta + \beta\gamma = 0$. This relation forces $\beta \neq 0$, $\delta \neq 0$ else the points S_{01}, S_{21} coincide; we can therefore assume that $S_{01} = (\alpha : 1 : 1)$, $S_{21} = (-\alpha : -1 : 1)$. There are two points left to consider. S_{10} lies on the line $x - z$ joining S_{00}, S_{20}; however, it also lies on the line $2x + (1 - \alpha)y - (\alpha + 1)z = 0$ joining S_{01}, S_{22}; thus

$S_{10} = (\alpha - 1 : \alpha + 3 : -\alpha + 1)$. Likewise S_{12} lies on the line $x = z$ joining S_{02}, S_{22}; however, it also lies on the line $2x - (\alpha + 1)y - (\alpha - 1)z = 0$ joining S_{01}, S_{20}; thus $S_{12} = (\alpha + 1 : -\alpha + 3 : \alpha + 1)$. Finally, the points S_{10}, S_{11}, S_{12} are collinear if and only if $\alpha^2 = -3$. It is now immediate that $S_{10} = (-1 : \alpha : 1)$ and $S_{12} = (1 : -\alpha : 1)$. □

It is worth pointing out that this is essentially a result over the complex field. Indeed the proof shows that *it is not possible to have a nine point configuration in* $\mathbb{PR}^2$ since the scalar α is necessarily non-real. Another corollary is that any nine point configuration in $\mathbb{PC}^2$ is projectively equivalent (in the above sense) to the configuration of flexes associated to the Steiner cubic. That has the following consequence.

Lemma 13.6 *Let F be any cubic curve in* $\mathbb{PC}^2$ *whose flexes form a nine point configuration. Then F is projectively equivalent to a Steiner cubic.*

Proof Since flexes are invariant under projective equivalences, F is projectively equivalent to a cubic G whose flexes form the nine point configuration of the Steiner cubics. And G is necessarily a Steiner cubic, by Lemma 13.4. □

In Chapter 17 we will see that the flexes of *any* non-singular cubic F in $\mathbb{PC}^2$ form a nine point configuration, and hence that any such cubic is projectively equivalent to a Steiner cubic. To obtain such information about the flexes we will produce a normal form (the Weierstrass normal form of Chapter 15) whose derivation depends crucially on the existence of a single flex, i.e. on the existence of at least one intersection of F and its Hessian H_F. That is why in the next chapter we address the general question of how two curves intersect.

14
Intersections of Projective Curves

In this central chapter, we justify (in some measure) the statement that the complex projective plane $\mathbb{PC}^2$ is the natural environment in which to study curves, by proving Bézout's Theorem: the number of intersections of two curves F, G in $\mathbb{PC}^2$ having no common component is the product of their degrees. Although the geometric idea behind the proof is compelling, the mechanics founder on a central difficulty, namely that of saying precisely what one means by the 'number of intersections' $I(P, F, G)$ of one curve F with another curve G at a point P. There are various ways forward here, each with its own merits and demerits. We will follow a classical geometric idea to arrive at a 'candidate' definition. The demerit of this approach is that the definition depends on a choice of coordinates, and it is by no means obvious that the result is independent of the choice. For the sake of completeness, we will present an invariance proof in the final section, based on standard ideas from complex analysis.

14.1 The Geometric Idea

We need the idea of 'projection' from a point onto a line in $\mathbb{PK}^2$. We are given a line L, and a point S not on L. Then for any point $Q \neq S$ there is a unique line M joining Q, S. Let Q_L to be the unique point where the line M meets L. (Figure 14.1.) We define *projection* of the plane, from the point S onto the line L, to be the mapping $Q \mapsto Q_L$.

Example 14.1 Suppose we have a coordinate system in which $S = (0 : 0 : 1)$ and L is the line $z = 0$. Take any point $Q = (a : b : c)$ distinct from S, then it is easily checked that the line joining Q, S is $bx - ay = 0$ and that it meets $z = 0$ at the point $Q_L = (a : b : 0)$. Thus, in this case projection is the mapping of the plane onto $z = 0$ defined by $(x : y : z) \mapsto (x : y : 0)$;

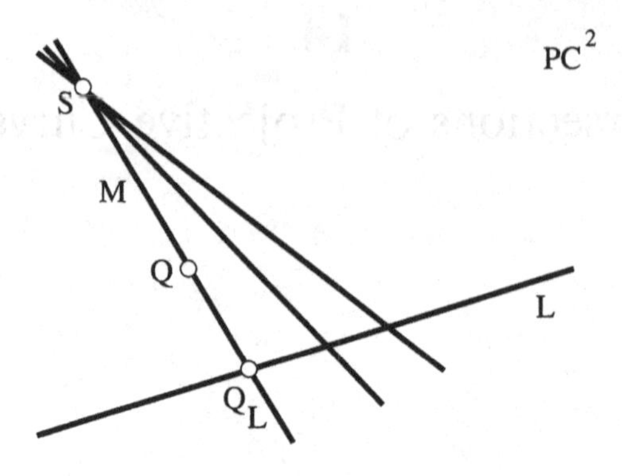

Fig. 14.1. Projection from a point to a line

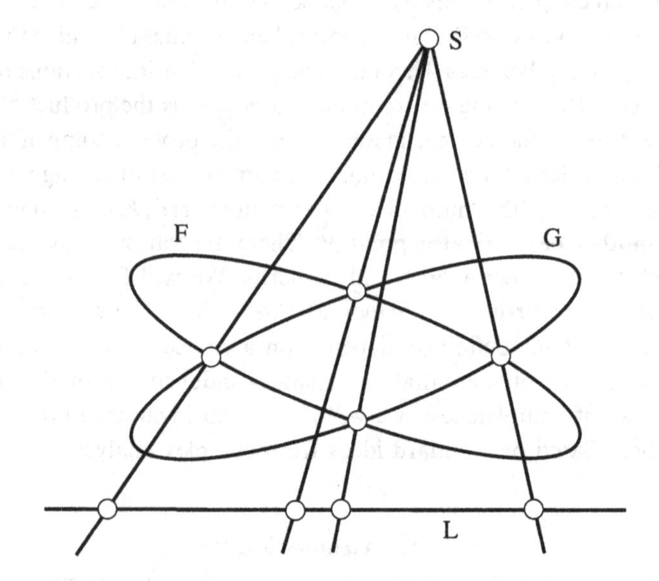

Fig. 14.2. Projecting intersections onto a line

think of it in the affine plane as 'outward' projection from the origin O onto a fixed circle centred at O.

The key geometric idea is as follows. Let F, G be curves in $\mathbb{PK}^2$ of respective degrees m, n having no common component. The object is to count their intersections. Choose a point S not on F or G, and project the plane from S onto some line L. (Figure 14.2.) What we intend to do is to count the points on L which arise from the intersections.

We can reduce the problem further. By the Four Point Lemma, we can assume $S = (0 : 0 : 1)$, and that L is the line $z = 0$. Think of F, G as polynomials in z, whose coefficients involve x, y by writing

$$\begin{cases} F &= F_0(x,y)z^m + \cdots + F_m(x,y) \\ G &= G_0(x,y)z^n + \cdots + G_n(x,y) \end{cases}$$

where the F_i, G_j are forms in x, y of respective degrees i, j. Now observe that $(x : y : 0)$ is the projection of an intersection of F, G if and only if there exists a z such that $F(x,y,z) = 0$, $G(x,y,z) = 0$. Think of x, y as fixed, so we require F, G (as polynomials in z) to have a common zero z. We need a technical result: that there exists a polynomial in the coefficients of F, G (as polynomials in z) giving a binary form $R(x,y)$ which vanishes if and only if F, G have a common zero. Then, in broad principle, the problem is reduced to that of counting the number of roots of $R(x,y)$. The key technical tool here is the 'resultant', whose study will occupy the next two sections.

14.2 Resultants in One Variable

This (and the next) section represents our second foray into the realm of algebra, in which we establish the basic properties of the 'resultant'. The starting point is the following result about polynomials in a single variable.

Lemma 14.1 *Let $f(t)$, $g(t)$ be polynomials of degrees p, q over a unique factorization domain $\mathbb{D}$. Then $f(t)$, $g(t)$ have a common non-constant factor if and only if there exist non-zero polynomials α, β of degrees $< p$, $< q$ with $f\beta = \alpha g$.*

Proof Suppose f, g have a common non-constant factor h, i.e. $f = h\alpha$, $g = h\beta$ for some non-zero polynomials α, β necessarily of degrees $< p$, $< q$, then $f\beta = \alpha g$. Conversely, suppose such α, β exist with $f\beta = \alpha g$. All these polynomials have unique factorizations into irreducibles. The factors of g (up to scalars) must appear amongst the factors of $f\beta$. They cannot all appear amongst the factors of β, since $\deg \beta < \deg g$, so at least one must be a factor of f, i.e. f, g have a common factor. □

The point of this lemma is that it can be converted into an entirely practical criterion. Let $f(t)$, $g(t)$ be polynomials over $\mathbb{D}$, defined by formulas

$$f(x) = a_o + a_1 t + \cdots + a_p t^p \qquad (a_p \neq 0)$$
$$g(x) = b_o + b_1 t + \cdots + b_q t^q \qquad (b_q \neq 0).$$

We define the *resultant* $R(f,g)$ to be the determinant of the following *resultant matrix* of order $(p+q)$, where there are q rows of a's and p rows of b's.

$$\begin{bmatrix} a_0 & a_1 & a_2 & \dots & a_p & 0 & 0 & \dots \\ 0 & a_0 & a_1 & a_2 & \dots & a_p & 0 & \dots \\ 0 & 0 & a_0 & a_1 & a_2 & \dots & a_p & \dots \\ \vdots & \vdots & \vdots & & \vdots & \vdots & \vdots & \\ b_0 & b_1 & b_2 & \dots & b_q & 0 & 0 & \dots \\ 0 & b_0 & b_1 & b_2 & \dots & b_q & 0 & \dots \\ 0 & 0 & b_0 & b_1 & b_2 & \dots & b_q & \dots \\ \vdots & \vdots & \vdots & & \vdots & \vdots & \vdots & \end{bmatrix}.$$

Lemma 14.2 *The polynomials f, g have a common non-constant factor if and only if $R(f,g) = 0$.*

Proof We show that $R(f,g) = 0$ if and only if there exist α, β as in Lemma 14.1. Set

$$\alpha(t) = \alpha_1 + \alpha_2 t + \cdots + \alpha_p t^{p-1} \qquad (\alpha_p \neq 0)$$
$$\beta(t) = \beta_1 + \beta_2 t + \cdots + \beta_q t^{q-1} \qquad (\beta_q \neq 0).$$

The relation $f\beta = \alpha g$ is then equivalent to the following linear system of $(p+q)$ equations in the $(p+q)$ unknowns $\alpha_1, \dots, \alpha_p, \beta_1, \dots, \beta_q$.

$$\begin{array}{lllcll} a_o\beta_1 & & & = & b_o\alpha_1 & \\ a_1\beta_1 + & a_o\beta_2 & & = & b_1\alpha_1 + & b_0\alpha_2 \\ \vdots & \vdots & \vdots & & \vdots & \vdots \quad \vdots \\ & & a_p\beta_q & = & & \quad b_q\alpha_p \end{array}$$

There exist non-zero polynomials α, β satisfying the relation $f\beta = \alpha g$ if and only if this system of equations has a non-trivial solution. By linear algebra that happens if and only if the determinant of the matrix of coefficients vanishes. Multiplying appropriate columns by -1, and transposing, we obtain the resultant matrix. The result follows. □

Example 14.2 Recall (Lemma 3.16) that a polynomial f in one variable t has a repeated root if and only if f, f' have a common zero. Thus f has a repeated root if and only if the *discriminant* $\Delta(f) = R(f, f') = 0$, a polynomial relation in the coefficients of f. For instance if $f(t) = at^2 + bt + c$ then $f'(t) = 2at + b$ and the discriminant is given by

$$\Delta(f) = \begin{vmatrix} c & b & a \\ b & 2a & 0 \\ 0 & b & 2a \end{vmatrix} = -a(b^2 - 4ac).$$

Thus the quadratic $f(t) = at^2 + bt + c$ with $a \neq 0$ has a repeated root if and only if $-a(b^2 - 4ac) = 0$ i.e. if and only if $b^2 - 4ac = 0$, a result familiar from school mathematics.

The resultant which appears in this example is very easily evaluated by hand. But in general the evaluation of a resultant is a decidedly non-trivial task. Fortunately, most computer algebra packages have a dedicated 'resultant' command allowing the resultant to be computed more or less instantaneously.

Exercises

14.2.1 Let $f(t) = t^3 + 4t - 1$, $g(t) = 2t^2 + 3t + 7$. Show that over the field $\mathbb{C}$ the resultant is non-zero, so f, g have no common zero. Show, however, that over the field $\mathbb{Z}_3$ the resultant is zero, and find the common zero.

14.2.2 Show that the cubic $t^3 + at + b$ has discriminant $\Delta = 4a^3 + 27b^2$.

14.2.3 In Example 2.9 we sketched the zero sets of the family of cubic curves $y^2 = \phi(x)$, where $\phi(x) = x^3 - 3x + \lambda$. Use the formula in the previous exercise to compute the discriminant of the polynomial $\phi(x)$, and deduce that $\phi(x)$ has a repeated factor if and only if $\lambda = \pm 2$. What are the repeated factors? Now correlate this information with the sketches in Figure 2.4.

14.2.4 Use the discriminant to show that the polynomial defined by $f(t) = 6t^4 - 23t^3 + 32t^2 - 19t + 4$ has a repeated root in $\mathbb{C}$.

14.3 Resultants in Several Variables

The main applications of resultants will be to the case when $\mathbb{D}$ is the domain $\mathbb{K}[x, y]$ of polynomials over a field $\mathbb{K}$ in the variables x, y. Here is a typical situation where one can profit from having set up the basic ideas over a general domain.

Example 14.3 We have come across a number of curves over a field $\mathbb{K}$ with rational parametrizations $x = x(t) = p(t)/q(t)$, $y = y(t) = r(t)/s(t)$

where $q \neq 0$, $s \neq 0$. We claim that the set of points in $\mathbb{K}^2$ of the form $(x(t), y(t))$ is *contained* in the zero set of a curve. To this end consider the polynomials $f(x, y, t) = xq(t) - p(t)$, $g(x, y, t) = ys(t) - r(t)$ over the unique factorization domain $\mathbb{K}[x, y]$, and write $R(x, y)$ for their resultant with respect to t. If (x, y) is a point in the image, there exists a t for which $f(t) = 0$, $g(t) = 0$ so the resultant $R(x, y) = 0$; thus the image is contained in the zero set of the polynomial $R(x, y)$. A special case arises when $x(t)$, $y(t)$ are polynomials of degrees d, e. In that case, one sees (by inspection) that $R(x, y)$ has degree $\max(d, e)$. Note that the argument does not establish that the set of points $(x(t), y(t))$ *coincides* with the zero set of $R(x, y)$, and in general that fails to be the case. (See Exercise 1.2.3 for a counterexample.)

Example 14.4 The Bézier curves of Section 1.2 are parametrized by polynomials of degree 3, so are contained in the zero set of a cubic curve. For instance the image of the curve $x(t) = t + t^2$, $y(t) = t^2 + t^3$ in Example 1.9 is contained in the zero set of the cubic curve $f = x^3 - xy - y^2$. The reader is invited to check that that agrees with the resultant of the polynomials $x - t - t^2$, $y - t^2 - t^3$ with respect to the variable t.

The following lemma is one of the technical keys to our discussion of intersection numbers.

Lemma 14.3 *Let F, G be polynomials in z of degrees m, n over the domain $\mathbb{K}[x, y]$ defined by formulas*

$$F(x, y, z) = A_m + A_{m-1}z + \cdots + A_0 z^m \qquad (A_0 \neq 0)$$
$$G(x, y, z) = B_n + B_{n-1}z + \cdots + B_0 z^n \qquad (B_0 \neq 0)$$

where A_k, B_k are forms in x, y of degree k. Then the resultant $R(x, y)$ of F, G with respect to z is either identically zero, or a form of degree mn.

Proof The resultant $R(x, y)$ is given by a determinant of order $(m + n)$. And since the A_k, B_k are forms of degree k the resultant $R(tx, ty)$ is obtained from it by multiplying each entry A_k, B_k by t^k. (We leave the reader to write out the determinant.) Multiply the first row of A's by t^n, the second by t^{n-1}, and so on till the n-th is multiplied by t. That multiplies the determinant by t^N, where $N = \frac{1}{2}n(n + 1)$, the sum of the first n natural numbers. Likewise, multiply the first row of B's by t^m, the second by t^{m-1}, and so on, till the m-th is multiplied by t. That multiplies the determinant by t^M where $M = \frac{1}{2}m(m + 1)$. Now take out a factor of

t^{m+n} from the first column, a factor of t^{m+n-1} from the second column, and so on till finally we take out a factor of t from the last column. That divides the determinant by t^W where $W = \frac{1}{2}(m+n)(m+n+1)$. The net result is to divide the determinant by $t^{W-N-M} = t^{mn}$. Thus $R(tx, ty) = t^{mn}R(x, y)$. The result follows from Lemma 3.12. □

Exercises

14.3.1 In each of the following cases, use resultants to find an equation $f(x, y) = 0$ for the curve with the given parametrization $x(t)$, $y(t)$.

(i) $x(t) = t^4, \ y(t) = t + t^2$
(ii) $x(t) = t^2 + t^3, \ y(t) = t^4$
(iii) $x(t) = t^2/(1+t^2), \ y(t) = t^3/(1+t^2)$.

14.4 Bézout's Theorem

The object of this section is to establish one of the most basic results in the theory of algebraic curves, namely Bézout's Theorem. We start with a 'weak' version of the result.

Lemma 14.4 *Let F, G be curves of degrees m, n in $P\mathbb{K}^2$ having no common component. Then F, G intersect in at most finitely many points. In fact, F, G intersect in $\leq mn$ distinct points.*

Proof The geometric idea was explained in Section 14.1 and illustrated in Figure 14.2. We will repeat the reasoning, filling in the technical detail. Choose a point S not on F or G, and project the plane from S onto some line L. On any line through S there are only finitely many intersection points of F, G, otherwise, that line is a common component of F, G. Thus it suffices to show there are only finitely many points on L which arise from intersections of F, G. By the Four Point Lemma we can assume $S = (0 : 0 : 1)$, and that L is the line $z = 0$. Set

$$\begin{cases} F &= F_0(x, y)z^m + \cdots + F_m(x, y) \\ G &= G_0(x, y)z^n + \cdots + G_n(x, y) \end{cases}$$

where the F_i, G_j are forms in x, y of respective degrees i, j; in particular, F_0, G_0 are constants, both $\neq 0$ as P does not lie on F or G. Then $(x : y : 0)$ is the projection of an intersection if and only if there exists a z such that $F(x, y, z) = 0$, $G(x, y, z) = 0$, i.e. if and only if F, G (as polynomials in z) have a common root z. By Lemma 14.2, there exists a polynomial $R(x, y)$ in the coefficients of F, G with the property that $F(x, y, z)$, $G(x, y, z)$ (as polynomials in z) have a common zero if and only if $R(x, y) = 0$. By Lemma 14.3, the resultant of F, G with respect to z is either zero, or a form of degree mn; we can exclude the former possibility, as F, G have no common factor. Thus the points $(x : y : 0)$ on L which arise from intersections of F, G correspond precisely to zeros $(x : y)$ of the binary form $R(x, y)$. Since there are only finitely many zeros $(x : y)$, we have established the first assertion. However, we need to argue more carefully to establish the second assertion; the trouble is that a ratio $(x : y)$ for which $R(x, y) = 0$ could correspond to *several* intersections $(x : y : z)$. We know there are only finitely many intersections, hence only finitely many lines joining them in pairs. Choose S not on any of these lines, and proceed as above. We now have a one-to-one correspondence between ratios $(x : y)$ with $R(x, y) = 0$, and intersections $(x : y : z)$. The result follows, as $R(x, y)$ has $\leq mn$ zeros. □

Even this relatively primitive result is useful. For instance the following application solves a problem raised in Example 2.11.

Lemma 14.5 *Two irreducible curves F, G in $P\mathbb{C}^2$ having no common component have the same zero set if and only if they coincide as curves.*

Proof Certainly, if F, G coincide as curves then their zero sets coincide. Conversely, suppose the zero sets coincide. We know that the zero set of any curve in $P\mathbb{C}^2$ is infinite, so it follows immediately from the previous lemma that F, G have a common component. Since F, G are irreducible that means they coincide as curves. □

The above considerations suggest a general definition of the term 'intersection number'. Let $F(x, y, z)$, $G(x, y, z)$ be curves in $P\mathbb{K}^2$, and let $P = (x, y, z)$ be any point in $P\mathbb{K}^2$. We will say that F, G intersect *properly* at P when F, G have no common component passing through P. Define $I(P, F, G) = \infty$ when P is a non-proper intersection of F, G, and to be $I(P, F, G) = 0$ when P is not an intersection. Now assume F, G have no common component. Choose coordinates X, Y, Z in which $(0 : 0 : 1)$ does not lie on $F(X, Y, Z)$, $G(X, Y, Z)$ or any of the lines joining their intersections, and let $R(X, Y)$ be the resultant with respect to Z. Then we can define the *intersection number* $I(P, F, G)$ for $P = (X : Y : Z)$ to be the multiplicity of $(X : Y)$ as a root of $R(X, Y)$. For the moment, we will assume a basic fact, namely that *the intersection number does not* *depend on the choice of coordinates X, Y, Z.* The interested reader will find a proof in the final section of this chapter. For consistency, we need to check that, in the case when L is a line, this definition is consistent with that of $I(P,F,L)$ in Section 10.1.

Lemma 14.6 *Let F be a curve of degree d in $P\mathbb{K}^2$, let L be a line which is not a component of F, and let P be a point, then the above definition of $I(P,F,L)$ agrees with that given in Section 10.1.*

Proof We are assuming intersection numbers are independent of the coordinate system, so (by the Four Point Lemma) can suppose that $P = (1:0:0)$, and that L is the line $z = 0$. Write

$$F(x,y,z) = F_d(x,y) + F_{d-1}(x,y)z + \cdots + F_0(x,y)z^d$$

where each F_k is a form of degree k. The resultant $R(x,y)$ of F, L with respect to the variable z is the diagonal matrix

$$\begin{vmatrix} F_d & F_{d-1} & \ldots & F_1 & F_0 \\ 0 & 1 & \ldots & 0 & 0 \\ \vdots & \vdots & \vdots & \vdots & \vdots \\ 0 & 0 & \ldots & 1 & 0 \\ 0 & 0 & \ldots & 0 & 1 \end{vmatrix}$$

Thus $R(x,y) = F_d(x,y)$ and $I(P,F,L)$, according to the above definition, is the multiplicity of $(1:0)$ as a zero of $F_d(x,y)$. According to the definition in Section 10.1 we find $I(P,F,L)$ by setting $z = 0$ in $F(x,y,z) = 0$, and then calculating the multiplicity of $(1:0)$ in the resulting binary form. But $F(x,y,0) = F_d(x,y)$, so the two definitions agree. □

Example 14.5 The parabolas $f = y - x^2 + 1$, $g = y - x^2 - 1$ do not intersect in $\mathbb{C}^2$. The corresponding conics in $P\mathbb{C}^2$ are $F = yz - x^2 + z^2$, $G = yz - x^2 - z^2$. F, G are irreducible (as they are clearly non-singular), do not pass through $(0:0:1)$, and meet $z = 0$ only at $P = (0:1:0)$. The resultant $R(x,y)$ of F, G with respect to z is

$$\begin{vmatrix} -x^2 & y & 1 & 0 \\ 0 & -x^2 & y & 1 \\ -x^2 & y & -1 & 0 \\ 0 & -x^2 & y & -1 \end{vmatrix} = 4x^4$$

so has a unique zero $(0:1)$ of multiplicity 4. Thus $I(P,F,G) = 4$.

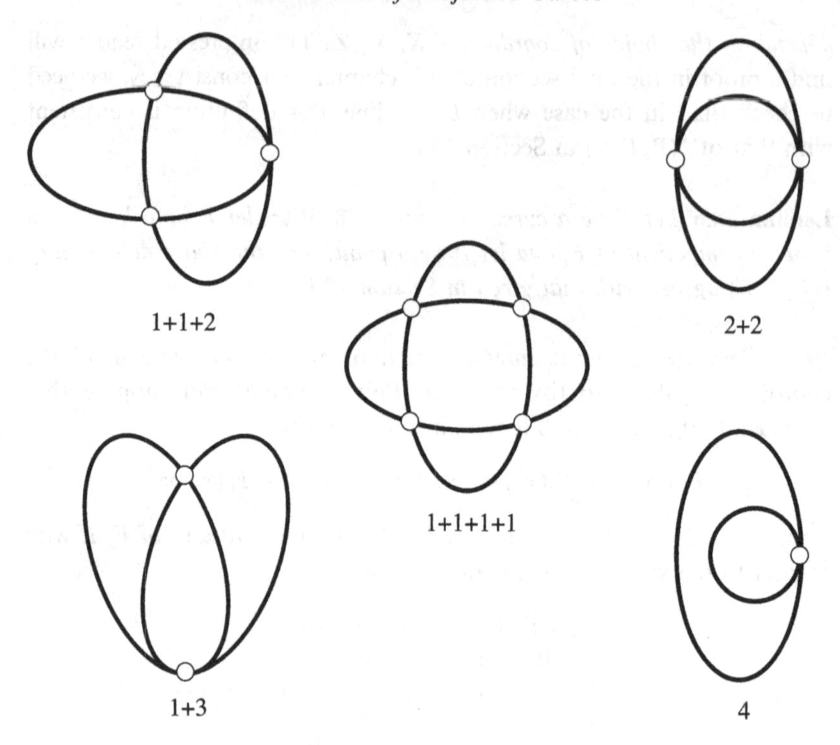

Fig. 14.3. Intersection patterns of two conics

Theorem 14.7 *Let F, G be curves in $\mathbb{PC}^2$ of degrees m, n with no common component. Then the sum of the intersection numbers $I(P,F,G)$ at the points of intersection P is mn. (We say that there are mn intersections, 'counted properly'.) In particular, F, G intersect in at least one point.* (Bézout's Theorem.)

Proof The sum of the multiplicities of the zeros of $R(x, y)$ is exactly mn, the degree of R. □

In view of the above we obtain the following picture of how a curve F of degree m in $\mathbb{PC}^2$ meets a curve G of degree n having no common component. There are $\leq mn$ distinct points of intersection $P_1, \ldots, P_s$ to which can be associated intersection numbers $i_1, \ldots, i_s$, where $i_k = I(P_k, F, G)$, with $i_1 + \cdots + i_s = mn$. This sum of positive integers is the *intersection pattern* of F with G, and is invariant under projective mappings. For instance, the possible intersection patterns for two conics correspond to the partitions of 4, namely $1 + 1 + 1 + 1$, $2 + 1 + 1$,

$2+2$, $3+1$ and 4. (Figure 14.3.) In fact all five possibilities can occur. (Exercise 14.4.5.)

Example 14.6 Any reducible curve F in $\mathbb{PC}^2$ is singular. (Hence, any non-singular curve in $\mathbb{PC}^2$ is irreducible.) Suppose $F = GH$ with G, H non-constant. We claim that G, H intersect in at least one point P, it is then immediate from Lemma 10.5 that P is singular on F. If G, H have a common component, then its zero set is infinite, and G, H certainly intersect in at least one point; otherwise, the claim follows from Bézout's Theorem.

Example 14.7 Any non-singular curve F of degree $d \geq 3$ in $\mathbb{PC}^2$ has at least one flex. Recall from Lemma 13.2 that the flexes of F are the simple points of intersection of F with its Hessian curve H_F, of degree $3(d-2)$. There are $3d(d-2)$ intersections counted properly, necessarily simple on F, and hence flexes. For instance a non-singular cubic in $\mathbb{PC}^2$ has nine flexes, counted properly. (In Lemma 15.3 we will see that they are distinct.)

Exercises

14.4.1 The curves F, G in $\mathbb{PC}^2$ are defined by $F = y^5 - x(y^2 - xz)^2$ and $G = y^4 + y^3z - x^2z^2$. Show that the resultant of these polynomials with respect to z is $R(x, y) = -x^3y^9(2x-y)^2$. Hence find the intersections of F, G and the intersection numbers at these points.

14.4.2 The curves F, G in $\mathbb{PC}^2$ are defined by $F = (x^2+y^2)^2+3x^2yz-y^3z$ and $G = (x^2+y^2)^3 - 4x^2y^2z^2$. Find the points of intersection of F, G and the intersection numbers at these points.

14.4.3 The curves F, G in $\mathbb{PC}^2$ are defined by $F = x^4 + y^4 - y^2z^2$ and $G = x^4 + y^4 - 2y^3z - 2x^2yz - xy^2z + y^2z^2$. Find the points of intersection of F, G and the intersection numbers at these points.

14.4.4 The curves F, G in $\mathbb{PC}^2$ are defined by $F = x^3 - y^3 - 2xyz$ and $G = 2x^3-4x^2y-3xy^2-y^3-2x^2z$. Find the points of intersection of F, G and the intersection numbers at these points.

14.4.5 The possible intersections of two irreducible conics in $\mathbb{PC}^2$ are represented by the possible partitions of 4, namely $1+1+1+1$, $2+1+1$, $2+2$, $3+1$ and 4. Show, by means of explicit examples, that all five types can occur.

14.4.6 Let F be an irreducible quartic curve in $\mathbb{PC}^2$. Show that F cannot have more than three singular points. (Consider the intersections of F with conics passing through all the singular points.)

14.5 The Multiplicity Inequality

Let F, G be curves in $\mathbb{PC}^2$ with no common component, and let P be a point of intersection, having multiplicity r on F, and s on G. Imagine the situation when there are r 'branches' of F, and s 'branches' of G, passing through P. Then one of the basic instincts is that each of the r 'branches' of F meets each of the s 'branches' of G at P, yielding at least rs intersections in total. The next result formalizes this instinct, and will prove to be a useful technical fact.

Lemma 14.8 *Let F, G be curves in $\mathbb{PC}^2$ with no common component, and let P be a point of intersection of multiplicities r, s respectively on F, G. Then $I(P,F,G) \geq rs$.* (The Multiplicity Inequality.)

Proof Let F, G have respective degrees m, n. We are working on the assumption (to be verified in the final section) that intersection numbers are invariant under projective changes of coordinates. Choose coordinates in which $(0:0:1)$ lies neither on F, G nor on any of the lines joining their intersections, and in which $P = (1:0:0)$. We wish to show that $(1:0)$ is a zero of the resultant of $R(F,G)$, with respect to z, of multiplicity $\geq rs$. Set $f(x,y) = F(x,y,1)$, $g(x,y) = G(x,y,z)$. Then the required multiplicity is that of the zero $x = 0$ of the resultant $R(f,g)$, with respect to y. Since P has multiplicities r, s on F, G we can write

$$\begin{aligned} f(x,y) &= \{f_0x^r + f_1x^{r-1}y + \cdots + f_ry^r\} + f_{r+1}y^{r+1} + \cdots \\ g(x,y) &= \{g_0x^s + g_1x^{s-1}y + \cdots + g_sy^s\} + g_{s+1}y^{s+1} + \cdots \end{aligned}$$

where $f_1, \ldots, f_r, g_1, \ldots, g_s$ are scalars, and $f_{r+1}, \ldots, g_{s+1}, \ldots$ are polynomials in x. Thus the resultant $R(f,g)$ of $f(x,y)$, $g(x,y)$ with respect to the variable y is

$$\begin{bmatrix} f_ox^r & f_1x^{r-1} & \ldots & f_{r+1} & \ldots & f_m & 0 & \ldots \\ 0 & f_ox^r & \ldots & f_r & \ldots & f_{m-1} & f_m & \ldots \\ \vdots & \vdots & & \vdots & & \vdots & \vdots & \vdots \\ g_ox^s & g_1x^{s-1} & \ldots & g_{s+1} & \ldots & g_n & 0 & \ldots \\ 0 & g_ox^s & \ldots & g_s & \ldots & g_{n-1} & g_n & \ldots \\ \vdots & \vdots & & \vdots & & \vdots & \vdots & \vdots \end{bmatrix}.$$

Multiply the first row of f's by x^s, the second by x^{s-1}, and so on till the s-th is multiplied by x. That multiplies the determinant by x^S where $S = \frac{1}{2}s(s+1)$, the sum of the first s natural numbers. Next, multiply the first row of g's by x^r, the second by x^{r-1}, and so on till the r-th is multiplied by x. That multiplies the determinant by x^R, where $R = \frac{1}{2}r(r+1)$. The result on the matrix is that the columns are multiplied (from left to right) by $x^{r+s}, x^{r+s-1}, \dots, x$. Dividing the columns by these values we divide the determinant by x^N where $N = \frac{1}{2}(r+s)(r+s-1)$. The net result on the determinant is to divide it by $x^{N-R-S} = x^{rs}$. Thus $R(f,g)$ has a factor x^{rs}, as was required. □

A more careful proof shows that we have a strict inequality $I(P,F,G) > rs$ in the Multiplicity Inequality if and only F, G are tangent at P; equivalently, $I(P,F,G) = rs$ if and only if F, G do not have a common tangent at P. In particular, $I(P,F,G) = 1$ if and only if P is simple both on F and on G, and the tangents at P are distinct. That yields a useful corollary to the Multiplicity Inequality, namely

Lemma 14.9 *Let F, G be curves in $\mathbb{PC}^2$ of degrees m, n with no common component, intersecting in mn distinct points. Then each point of intersection P is simple on F and on G, and the tangents at P are distinct.*

The general picture is that even if two curves of degrees m, n fail to intersect in mn distinct points, that can always be achieved by arbitrary small deformations of one of the equations. A proof of this fact requires topological concepts, so we will content ourselves with an example.

Example 14.8 Consider the conics $F_\delta = x^2 + y^2 + (\delta^2 - 1)z^2$, $G = x^2 + 2y^2 - z^2$ in $\mathbb{PR}^2$ where $\delta \geq 0$ is small. (In the affine view $z = 1$ they represent respectively a small deformation of the unit circle, and an ellipse.) The reader is left to check that when $\delta = 0$ the curves F_δ, G intersect in just two distinct points (with both intersection numbers equal to 2) but that when $\delta > 0$ they intersect in four distinct points.

Here is a typical application of the Multiplicity Inequality, producing bounds for the number (and for the multiplicities) of singular points on a projective curve.

Lemma 14.10 *Let F be an irreducible curve of degree d in $\mathbb{PC}^2$. Then F has only finitely many singular points. Indeed, if $P_1, \dots, P_s$ are the singular points, of multiplicities $m_1, \dots, m_s$, then*

$$\sum_{k=1}^{s} m_k(m_k - 1) \leq d(d-1).$$

Proof Note first that the curves F, F_z have no common component, since F is irreducible (of degree d) and F_z has degree $(d-1)$. By Bézout's Theorem, they intersect in $d(d-1)$ points, counted properly. A singular point P of F is necessarily an intersection point of F, F_z so there are at most finitely many. Further, if X has multiplicity m on F then X has multiplicity $\geq (m-1)$ on F_z, in view of Lemma 6.5. The result now follows from the following inequalities, where S denotes the set of intersections of F, F_z.

$$d(d-1) = \sum_{P\in S} I(P,F,F_z) \geq \sum_{k=1}^{s} I(P_k,F,F_z) \geq \sum_{k=1}^{s} m_k(m_k-1).$$

□

As it stands, Lemma 14.10 is not particularly useful. For instance, for irreducible cubics ($d = 3$) it only tells us that there are at most *three* singular points. Using the idea of a 'linear system' of curves, we will derive better estimates in Section 18.3.

Exercises

14.5.1 The curves F, G in $P\mathbb{C}^2$ are defined by $F = y^2z - x(x-2z)(x+z)$ and $G = x^2 + y^2 - 2xz$. Show that F, G intersect in two real points, and two complex conjugate points, and are tangent at both of the real points. What are the intersection numbers of F, G at those points?

14.5.2 Find the intersections of the conic $F = y^2 - xz$ and the cubic $G = y^2z - xz^2 + x^3$ in $P\mathbb{C}^2$, and the intersection numbers at these points.

14.5.3 Find the intersections of the cubics $F = (x^2 + y^2)z + x^3 + y^3$, $G = x^3 + y^3 - 2xyz$ in $P\mathbb{C}^2$, and the intersection numbers at these points.

14.5.4 Show that the curves $F = (x^2 + y^2)^2 - 2yz(x^2 + y^2) - x^2z^2$ and $G = x^2 + y^2 - yz$ in $P\mathbb{C}^2$ have exactly three distinct intersections, namely the circular points at infinity I, J and $O = (0 : 0 : 1)$. Show that I, J, O are cusps on F, and find the three cuspidal tangents. Find the tangents to G at I, J, O. Determine the intersection numbers of F, G at I, J, O. (In the real affine view $z = 1$, the curve F is a cardioid, and G is a circle.)

14.5.5 Show that a cubic curve F in $P\mathbb{C}^2$ with exactly one singular point, namely a node, is necessarily irreducible. Now extend the argument to curves of degree ≥ 3.

14.5.6 Show that a quartic curve F in $P\mathbb{C}^2$ with exactly one singular point P, namely an ordinary triple point, is necessarily irreducible. Now extend the argument to curves of degree ≥ 4. Show, by means of an example, that the conclusion may fail if the hypothesis that P is ordinary is omitted.

14.5.7 Let Q be a quartic curve in $P\mathbb{C}^2$ with exactly three non-collinear singular points, which are all nodes. By considering the possible ways in which F might reduce, show that F must be irreducible. Use the results of this question to show that the quartic curve $Q = x^2y^2 + y^2z^2 + z^2x^2$ in $P\mathbb{C}^2$ is irreducible.

14.5.8 In each of the following cases give an example of a *reducible* quartic curve F in $P\mathbb{C}^2$ with the given properties. (i) F has exactly one singular point, namely a cusp. (ii) F has exactly two singular points, one a cusp and one a node. (iii) F has exactly two singular points, both cusps. (iv) F has exactly three non-collinear singular points, one a cusp and two nodes. (v) F has exactly three non-collinear singular points, one a node and two cusps.

14.5.9 Let Q be a quartic curve in $P\mathbb{C}^2$ with exactly three non-collinear singular points, which are all cusps. By considering the possible ways in which F might reduce, show that F must be irreducible.

14.6 Invariance of the Intersection Number

Here is the promised proof of the invariance of the intersection number $I(P, F, G)$ under projective changes of coordinates, taken from *Plane Algebraic Curves* by E. Brieskorn and H. Knörrer. It uses results from complex analysis, and is included for the sake of completeness. The reader can omit the proof without jeopardizing future understanding.

Lemma 14.11 *Let F, G be curves in $P\mathbb{C}^2$ of degrees m, n having no common component. The intersection number $I(P, F, G)$ defined above is invariant under projective changes of coordinates.* (Invariance of the Intersection Number.)

Proof For clarity, we will split the proof into several steps, representing different lines of thought.

Step 1: F, G have only finitely many intersections, so there are only finitely many lines $L_1, \dots, L_v$ joining them in pairs. Write $H = FGL_1 \dots L_v$. Thus the zero set of H comprises the union of the zero sets of F, G and the lines joining their intersections. A change from coordinates X, Y, Z to the standard coordinates x, y, z is identified with an invertible 3×3 matrix $A = (a_{ij})$, and maps $(0:0:1)$ to a point $P(A)$. Write M for the 9–dimensional affine space of complex 3×3 matrices, and define polynomials U, V, W on M by $U(A) = H(P(A))$, $V(A) = \det A$ and $W = UV$. Thus $W(A) \neq 0$ means that A represents a change of coordinates, and that $(0:0:1)$ does not lie on the curves $F(X,Y,Z)$, $G(X,Y,Z)$ or any of the lines joining their intersections. These are the only matrices A in M which concern us.

Step 2: We claim that for *any* non-zero polynomial W on M, the set of matrices X with $W(X) \neq 0$ is *path connected*. This means that given any two such matrices A, A' there is a continuous path of matrices A_t in M $(0 \leq t \leq 1)$ with $A = A_0$, $A' = A_1$ and $W(A_t) \neq 0$ for all t. To this end, consider the parametrized affine line $(1-z)A + zA'$ joining A, A' in M where z is a complex number. Then $w(z) = W((1-z)A + zA')$ is a non-zero complex polynomial in z, so has only finitely many zeros. The claim follows on observing that there exists a continuous path z_t $(0 \leq t \leq 1)$ in the complex plane with $0 = z_0$, $1 = z_1$ which avoids the zeros of $\theta(z)$, and then defining $A_t = (1 - z_t)A + z_tA'$. (We are simply noting that any two distinct points in the complex plane can be joined by a path avoiding finitely many other given points.)

Step 3: We will use the following fact. Let $r_t(x)$ be a family of complex polynomials of constant degree d, whose coefficients depend continuously on the real variable t. It is assumed that the number of *distinct* zeros of $r_t(x)$ takes a constant value. Then the multiplicities of the zeros also take constant values. The statement is an immediate consequence of the following standard result in complex analysis. (A corollary of the so-called Principle of the Argument.) Each zero of multiplicity μ of a complex polynomial $f(x)$ has a neighbourhood N in the complex plane, with the property that the sum of the multiplicities of the zeros of $g(x)$ in N takes the constant value μ for any complex polynomial $g(x)$ sufficiently close to $f(x)$. The statement extends immediately to any family $R_t(x, y)$ of complex binary forms.

Step 4: Let A be a matrix in M with $W(A) \neq 0$ yielding coordinates X, Y, Z, and A' another with $W(A') \neq 0$ yielding coordinates X', Y', Z'. By Step 2, there is a continuous path A_t in M with $A = A_0$, $A' = A_1$ and $W(A_t) \neq 0$ for all t. A_t corresponds to coordinates X_t, Y_t, Z_t. Let

$R(X_t, Y_t)$ be the resultant of $F(X_t, Y_t, Z_t)$, $G(X_t, Y_t, Z_t)$ with respect to Z_t. Then, since X_t, Y_t, Z_t depend continously on the standard coordinates x, y, we see that $R_t(x, y) = R(X_t, Y_t)$ is a family of complex binary forms of constant degree mn, whose coefficients depend continuously on t. The number of distinct zeros of $R_t(x, y)$ is constant, namely the number of distinct intersections of F, G. By Step 3, the multiplicities of the zeros are likewise constant. In particular, the multiplicities of $R_0(x, y)$, $R_1(x, y)$ are equal, i.e. the intersection numbers computed relative to the coordinates X, Y, Z and X', Y', Z' are equal. □

15
Projective Cubics

Our attempts at obtaining complete lists of curves of a given degree have been somewhat limited – apart from lines, we have only managed to list conics. However, it is curves of higher degree which proliferate in applications, and it is time for us to make progress on the next case of cubics. However, we have learnt important lessons from our studies of conics. It is easier to list in the complex case than the real case, and easier to list in the projective than the affine case. On that basis, the natural way forward is to attempt to list complex projective cubics. We will approach this classification in two steps. First, we will use the basic geometry of cubics to categorize (complex projective) cubics into nine distinct 'geometric types': that is the object of Section 15.1. Then in the remainder of the chapter we will show that our categorization is *almost* a classification up to projective equivalence.

15.1 Geometric Types of Cubics

The basic fact which allows us to categorize cubic curves in $P\mathbb{C}^2$ into a small number of types is the following lemma.

Lemma 15.1 *An irreducible cubic F in $P\mathbb{C}^2$ has at most one singular point, necessarily of multiplicity 2, so either a node or a cusp.*

Proof F cannot have two distinct singular points P_1, P_2; for then the line L joining P_1, P_2 would meet F in $\geq 2+2=4$ points, so have to be a component of F, contradicting irreducibility. Suppose F has a singular point P of multiplicity ≥ 3. Let Q be any point on F distinct from P. (Recall that the zero set of F is infinite, so there certainly exists at least one such point Q.) Then the line L joining P, Q meets F in ≥ 4 points, so must be a component of F, contradicting irreducibility again. □

We can categorize cubic curves F in $P\mathbb{C}^2$ into nine *geometric types*. Consider first the case when F is irreducible. Either F is non-singular, or by Lemma 15.1 has exactly one singular point, necessarily a node or a cusp. A reducible cubic F reduces either to a line and an irreducible conic, or to three lines. In the former case the line either meets the conic in two distinct points, or is tangent to the conic. In the latter case, the three lines either form a triangle or are concurrent. And in the concurrent case we can have three distinct lines, or two distinct lines of which one is repeated, or just one line which is thrice repeated. Thus we arrive at the following list.

1.	*general*	non-singular, necessarily irreducible.
2.	*nodal*	irreducible, with a node.
3.	*cuspidal*	irreducible, with a cusp.
4.	*conic plus chord*	line not tangent to the conic.
5.	*conic plus tangent*	line is tangent to the conic.
6.	*triangle*	three non-concurrent lines.
7.	*three lines*	three distinct concurrent lines.
8.	*two lines*	two distinct lines, one of multiplicity 2.
9.	*one line*	one line of multiplicity 3.

Table 15.1 gives an example of each geometric type, followed by the number N of singular points, and the types of singular points which arise. (For the general type, there are no singularity types, and the relation $\alpha^3 \neq 27\beta^2$ is the condition for the curve to be non-singular.) Looking at the last two columns, we see that all but two of the geometric types are distinguished (up to projective equivalence) by their singularities. The cuspidal and conic plus tangent types fail to be distinguished by their singularities, but are projectively inequivalent since the former is irreducible, whilst the latter is reducible. Note that in the conic plus tangent type the unique tangent to the curve at the cusp is a component of the curve. Thus, given a cubic with just one singular point, namely a cusp, we can determine its geometric type by checking whether the tangent is a component.

Example 15.1 The reader is invited to check that the cubic in $P\mathbb{C}^2$ given by $F = x^3 + x^2y + xy^2 + y^3 - 2x^2z - 4xyz - 2y^2z$ has a unique singular point of multiplicity 2, at the point $P = (0 : 0 : 1)$, and that the tangent

Table 15.1. *Geometric types of cubics*

example	description	N	singularity types
$y^2z = 4x^3 - \alpha xz^2 - \beta z^3$	general type	0	$\alpha^3 \neq 27\beta^2$
$x^3 + y^3 - xyz$	nodal	1	node
$y^2z - x^3$	cuspidal	1	cusp
$x(x^2 + y^2 + z^2)$	conic plus chord	2	nodes
$x(x^2 - 2xz + y^2)$	conic plus tangent	1	cusp
$x(y^2 + z^2)$	triangle	3	nodes
$x(x^2 - y^2)$	three line type	1	ordinary triple point
x^2y	two line type	∞	triple point and cusps
x^3	one line type	∞	triple points

at P is the repeated line $(x + y)^2$. Thus P is a cusp, and F is of cuspidal or conic plus tangent type. However $F(x, -x, z)$ is identically zero, so the tangent at P is a component of F, and F must be of conic plus tangent type. In fact $F = (x + y)(x^2 + y^2 - 2xz - 2yz)$, with $x + y$ the tangent at P to the irreducible conic $x^2 + y^2 - 2xz - 2yz$.

It turns out that Table 15.1 is *almost* a classification up to projective equivalence; more precisely, we will show that if two *singular* cubics have the same geometric type then they are projectively equivalent. Roughly speaking, the more degenerate the curve, the easier it is to find a 'normal form'. We will start with the general type, and work our way through the singular irreducible curves, dealing finally with the reducible curves. The situation for general cubics turns out to be more complex, in that there are infinitely many projective equivalence classes.

Exercises

15.1.1 Verify that the singular cubics in Table 15.1 have the numbers and types of singularities indicated.

15.1.2 In each of the following cases, find the geometric type of the given cubic curve F in $\mathbb{PC}^2$.

(i) $F = x^3 + x^2z + y^2z$
(ii) $F = x^3 - xy^2 - yz^2$
(iii) $F = x^2z - y^2z + x^3 + 2y^3$
(iv) $F = y^2z + yz^2 + x^3 + y^3$
(v) $F = x^3 + y^3 + 3x^2y + 3xy^2 - xz^2 - yz^2$.

15.1.3 In each of the following cases, find the geometric type of the given cubic curve F in $P\mathbb{C}^2$. (a, b are positive reals.)

(i) $F = x^2y - 4a^2z^2(2az - y)$ (Agnesi's versiera)
(ii) $F = x^2y + a^2yz^2 - b^2xz^2$ (Serpentine)

15.1.4 Recall from Example 1.8 that the pedal curve of a standard parabola $y^2 = 4ax$ ($a > 0$) with respect to a pedal point $(\alpha, 0)$ on the x-axis is the cubic curve $x(x-\alpha)^2 + y^2(x-\alpha) + ay^2 = 0$. Assuming that the pedal has a double point at $(\alpha, 0)$ show that the corresponding curve in $P\mathbb{C}^2$ is reducible if and only if the pedal point is the focus $F = (a, 0)$ of the parabola; and in that case, show that the pedal reduces to three distinct lines.

15.1.5 A cubic curve F in $P\mathbb{C}^2$ is defined by $F_\lambda = (x+y+z)^3 + \lambda xyz$ where $\lambda \neq 0$ is a complex number. Show that F_λ is singular if and only if $\lambda = -27$. Determine the projective type of F_λ when $\lambda = -27$.

15.1.6 $F_k = x^3 + y^3 + z^3 + k(x+y+z)^3$ is a cubic curve in $P\mathbb{C}^2$, where k is a complex number. Show that F_k is non-singular, save for two exceptional values of k. What type of cubic is F_k for these exceptional values?

15.1.7 Show that there are exactly two values of the complex number λ for which the cubic curve $F = x^3 + y^3 + xyz + \lambda z^3$ in $P\mathbb{C}^2$ is singular. What type of cubic is F for these exceptional values of λ?

15.1.8 Show that the cubic curve $F = xy^2 + yz^2 + zx^2 + \lambda xyz$ in $P\mathbb{C}^2$ is non-singular save for three exceptional values of λ, when the curve is nodal.

15.1.9 $F = x^3 + xyz + Ay^3 + Bz^3$ is a cubic curve in $P\mathbb{C}^2$, where A, B are complex numbers. On the assumption that A, B are sufficiently small (in absolute value), show that F is of conic and chord type if and only if both A, B are zero, is nodal if and only if exactly onc of A, B is zero, and is non-singular if and and only if neither of A, B is zero.

15.1.10 $F = xyz + Ax^3 + By^3 + Cz^3$ is a cubic curve in $P\mathbb{C}^2$, where A, B, C are complex numbers. Show that if A, B, C are sufficiently small (for instance if their moduli are all $< 1/3$) then F is non-singular if and only if all of A, B, C are $\neq 0$, is nodal if and only if exactly one of A, B, C is zero, is of conic and chord type if and only if exactly two of A, B, C are zero, and is of triangle type if and only if all of A, B, C are zero.

15.2 Cubics of General Type

By a *Weierstrass normal form* (WNF) we mean a cubic in $\mathbb{PC}^2$ defined by an equation of the form $y^2z = \Phi(x, z)$ with $\Phi(x, z)$ a non-zero binary cubic form in x, z, thus in the affine view $z = 1$ such a curve has the form $y^2 = \phi(x)$ with $\phi(x)$ a cubic in x. In this section, we will establish that a general cubic curve in $\mathbb{PC}^2$ is projectively equivalent to a WNF, and use this fact to show that any general cubic has nine distinct flexes.

Lemma 15.2 *Any general cubic curve F in $\mathbb{PC}^2$ is projectively equivalent to a cubic $y^2z = 4x^3 - \alpha xz^2 - \beta z^3$ for some complex numbers α, β with $\alpha^3 \neq 27\beta^2$.*

Proof By previous results, F has a flex P. By the Four Point Lemma, we can assume $P = (0 : 1 : 0)$, and that the tangent to F at P is $z = 0$. These assumptions simplify the equation. The curve F is a linear combination of the monomials x^3, y^3, z^3, x^2y, x^2z, y^2x, y^2z, z^2x, z^2y, xyz. The condition for F to pass through P is that the coefficient of y^3 vanishes. The condition for $z = 0$ to be an inflexional tangent to F at P is that $F(x, y, 0)$, which is a linear combination of x^3, x^2y, y^2x, should have the ratio $(0 : 1)$ as a root of multiplicity 3, i.e. that the coefficients of x^2y, y^2x vanish. Thus F has the form

$$F = ax^3 + bx^2z + cxyz + dy^2z + ez^2x + fz^2y + gz^3.$$

Note first that $a \neq 0$, else F has a factor z. Calculation verifies that $F_x(P) = 0$, $F_y(P) = 0$, $F_z(P) = d$. And since P is non-singular $d \neq 0$. Replace y by $y - (c/2d)x - (f/2d)z$; that forces the coefficients of xyz, yz^2 to vanish, and does not re-introduce any of the monomials so far excluded. Having done that, replace x by $x - (b/3a)z$; that forces the coefficient of zx^2 to vanish, and likewise does not re-introduce any monomials so far excluded. Thus our cubic can be assumed to have the form $F = ax^3 + dy^2z + ez^2x + gz^3$. We can assume $d = 1$, by multiplying through by $1/d$. And we can assume $a = -4$ by scaling x. That yields the required WNF: it remains to show that the WNF is non-singular if and only if $\alpha^3 \neq 27\beta^2$. The WNF meets the line $z = 0$ only at the point P, so any singular points appear in the affine view $z = 1$. By Exercise 6.3.9, this affine curve $y^2 = 4x^3 - \alpha x - \beta$ is singular if and only if the cubic $4x^3 - \alpha x - \beta$ has a repeated root, and by Exercise 14.2.2, that is the case if and only if $\alpha^3 = 27\beta^2$, yielding the result. □

The question arises as to when two normal forms for general cubics are projectively equivalent. We will simply state the result. The function $j = \alpha^3/(\alpha^3 - 27\beta^2)$ turns out to be a projective invariant of the normal form, and two normal forms $y^2z = 4x^3 - \alpha xz^2 - \beta z^3$ are projectively equivalent if and only if the values of j coincide. In particular, there are infinitely many types of general cubic.

Example 15.2 An 'alternative' WNF for a general cubic in $\mathbb{PC}^2$ is $y^2z = x^3 + Ax^2z + Bxz^2$, obtained from $y^2z = 4x^3 - \alpha xz^2 - \beta z^3$ by replacing x by $x + \lambda$, and then choosing λ so that the constant term vanishes. (The constant term is a polynomial of degree 3 in λ so has at least one complex zero.) It is easily checked (Exercise 15.2.3) that the alternative WNF is general if and only if $B(A^2 - 4B) \neq 0$.

The virtue of having normal forms for a general cubic is that we can access geometric information by computations on the normal form. Here is a typical application of this philosophy. We saw in Example 14.7 that Bézout's Theorem allows us to deduce that a general cubic in $\mathbb{PC}^2$ has at most nine flexes. In fact more is true.

Lemma 15.3 *Any general cubic in* $\mathbb{PC}^2$ *has nine distinct flexes.*

Proof It suffices to work with the 'alternative' WNF of Example 15.2 given by $F = y^2z - x^3 - Ax^2z - Bxz^2$. (Recall that F is general if and only if $B(A^2 - 4B) \neq 0$.) The flexes of F are its intersections with the Hessian H_F. The reader is left to check that the Hessian is given by $H_F = (y^2 + Bxz)(3x + Az) - z(Ax + Bz)^2$. Clearly, the only intersection on the line $z = 0$ is the flex $(0 : 1 : 0)$. It suffices therefore to show that the affine views $f = y^2 - x^3 - Ax^2 - Bx$ and $h_f = (y^2 + Bx)(3x + A) - (Ax + B)^2$ have eight distinct intersections in the affine plane $z = 1$. Eliminating y between the relations $f = 0$, $h_f = 0$ we obtain a polynomial $k(x) = 3x^4 + 4Ax^3 + 6Bx^2 - B^2$ of degree 4. The polynomial $k(x)$ has four distinct roots, for its derivative $k'(x) = 12(x^3 + Ax^2 + Bx)$, and (by direct computation) the resultant of k, k' is $B^4(A^2 - 4B)^2$, which is $\neq 0$, by the initial remark. For each of these four values of x there are exactly two values of y. (None of them can satisfy the equation $x^3 + Ax^2 + Bx = 0$.) □

Exercises

15.2.1 Let F be a real non-singular cubic curve in $\mathbb{PC}^2$. Show that F has at least one *real* flex. (You may assume the results of Exercise 9.5.6 and of Lemma 15.3.)

15.2.2 Show that the cubic curve F_k in $\mathbb{PC}^2$ given by the following formula is general unless $k = -6$, $k = 2$ or $k = 3$.

$$F_k = xy^2 + yz^2 + zx^2 + x^2y + y^2z + z^2x + kxyz.$$

15.2.3 Show that the cubic curve $y^2z = x^3 + Ax^2z + Bxz^2$ is general if and only if $B(A^2 - 4B) \neq 0$.

15.2.4 Let F be a non-singular cubic curve in $\mathbb{PC}^2$. It can be shown that there are points X, Y, Z on F such that the tangent lines to F at X, Y, Z meet the curve again at Y, Z, X respectively. Show that F is projectively equivalent to a curve $ax^2z + bz^2y + cy^2x + dxyz$ for some complex numbers a, b, c, d with $abc \neq 0$. (Take X, Y, Z to be the vertices of reference.) Deduce that F is projectively equivalent to a curve $x^2z + z^2y + y^2x + \lambda xyz$, for some complex number λ.

15.3 Singular Irreducible Cubics

Lemma 15.4 *Any nodal cubic F in $\mathbb{PC}^2$ is projectively equivalent to the normal form $x^3 + y^3 = xyz$. (And hence any two nodal cubics in $\mathbb{PC}^2$ are projectively equivalent.)*

Proof F has a node P. By the Four Point Lemma, we can assume $P = (0 : 0 : 1)$ and that the tangents at P are $x = 0$, $y = 0$. The cubic F is a linear combination of x^3, y^3, z^3, x^2y, x^2z, y^2x, y^2z, z^2x, z^2y, xyz and passes through P if and only if the coefficient of z^3 vanishes. As P is singular, F_x, F_y, F_z vanish at P. Note that F_x, F_y vanish at P if and only if the coefficients of z^2x, z^2y vanish. (F_z vanishes automatically at P.) The lines $x = 0$, $y = 0$ must meet F at P with multiplicity 3. Setting $x = 0$ in F we get a linear combination of y^3, y^2z with $(0 : 1)$ as a root of multiplicity 3 if and only if the coefficient of y^2z vanishes. And setting $y = 0$ in F we get a linear combination of x^3, x^2z with $(0 : 1)$ as a root of multiplicity 3 if and only if the coefficient of x^2z vanishes. Thus $F = ax^3 + by^3 + cx^2y + dxy^2 + exyz$. Note that a, b, e are all non-zero, otherwise F reduces. We can suppose $a = 1$ (by dividing through by a) and that $b = 1$, $e = -1$ by scaling y, z. Finally, replace z by $z + cx + dy$ (a projective transformation) to get the required normal form. □

Lemma 15.5 *Any cuspidal cubic F in $\mathbb{PC}^2$ is projectively equivalent to the normal form $y^2z = x^3$. (And hence any two cuspidal cubics in $\mathbb{PC}^2$ are projectively equivalent.)*

Proof F has a cusp P. By the Four Point Lemma, we can suppose $P = (0:0:1)$ and that the tangent to F at P is $y = 0$ (repeated). As in the nodal case, P is singular on F if and only if the coefficients of z^3, z^2x, z^2y vanish. The line $y = 0$ meets F at P with multiplicity 3 if and only if the coefficients of x^2z, xyz vanish. Thus $F = ax^3+by^3+cx^2y+dxy^2+ey^2z$ for some a, b, c, d, e. The coefficients a, e are non-zero, else F reduces. We can suppose $a = 1$ (by dividing through by a) and that $e = -1$ (by scaling z). Replacing x by $x - \frac{c}{3}y$, we force the coefficient of x^2y to vanish. And replacing z by $z - dx - by$ we force the coefficients of y^3, xy^2 to vanish. That yields the desired normal form. □

We can deduce (projectively invariant) facts about the geometry of singular irreducible cubics by doing explicit calculations on the normal form. Indeed this provides an excellent motivation for the process of listing. For instance, in Example 13.3 it was shown that the normal form for a nodal cubic has exactly three collinear flexes, thus *any* nodal cubic in $\mathbb{PC}^2$ will have exactly three collinear flexes. (The exercises in Chapter 13 provide considerable experimental evidence for this fact.) Likewise, in Example 13.2 it was shown that the normal form for a cuspidal cubic has exactly one flex, thus *any* cuspidal cubic in $\mathbb{PC}^2$ will have exactly one flex. We will be able to throw some light on this in Chapter 17.

You get a good mental picture of irreducible cubics by considering the family of cubics $y^2z = 4x^3 - \alpha xz^2 - \beta z^3$ with *arbitrary* coefficients α, β. Think of this as a family of deformations of the cuspidal type $y^2z = 4x^3$ with the coefficients α, β varying. The cubic is general if and only if $\alpha^3 \neq 27\beta^2$, is nodal if and only if $\alpha^3 = 27\beta^2$ with at least one of α, β non-zero, and is cuspidal if and only if $\alpha = 0$, $\beta = 0$. For *real* values of α, β the affine curve $y^2 = 4x^3 - \alpha x - \beta$ can be sketched by tracing the graph of the cubic in x, and 'taking square roots'. You can see how the picture varies with α, β by moving around the origin in the (α, β)-plane, sketching as you go. (Figure 15.1.) The two parts of the curve $\alpha^3 = 27\beta^2$ (with the origin deleted) correspond to the crunodal and acnodal types; the region in between corresponds to general cubics in two bits, and the rest of the plane corresponds to general cubics in one bit.

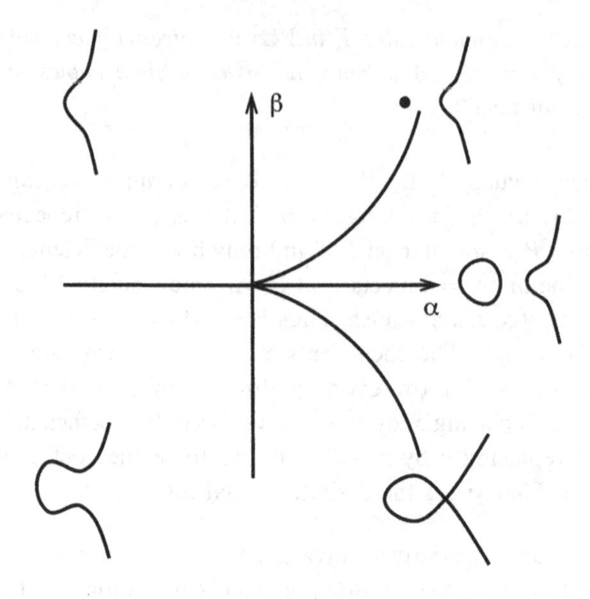

Fig. 15.1. Deforming a cuspidal cubic

Exercises

15.3.1 Show that any cuspidal cubic F in the real projective plane $P\mathbb{R}^2$ is projectively equivalent to the normal form $y^2z = x^3$. (And hence any two cuspidal cubics in $P\mathbb{R}^2$ are projectively equivalent.)

15.3.2 Let F be an irreducible cubic curve in $P\mathbb{C}^2$ with a node at $P = (1 : 0 : 0)$ having nodal tangents $y = 0$, $z = 0$. Show that F has the form $F = (ax + by + cz)yz + (dy + ez)^3$ with a, d, e all $\neq 0$. Deduce that F is projectively equivalent to the curve $xyz + (y + z)^3$. Hence show that any nodal cubic curve in $P\mathbb{C}^2$ is projectively equivalent to this normal form. (Mimic the derivation of the normal form $x^3 + y^3 + xyz$ for nodal cubics, but do not appeal to the fact that it is a normal form.)

15.4 Reducible Cubics

Lemma 15.6 *Any two cubic curves in* $P\mathbb{C}^2$ *of conic plus chord type, or of conic plus tangent type, are projectively equivalent.*

Proof In order to make the proof as transparent as possible we will split it into a number of steps.

Step 1: We claim first that if A_1, A_2, A_3 are distinct points on an irreducible conic F in $\mathbb{PC}^2$ then there is a projective map taking A_1, A_2, A_3 respectively to $E_1 = (1 : 0 : 0)$, $E_2 = (0 : 1 : 0)$, $E_3 = (0 : 0 : 1)$, and F to the conic $xy + yz + zx$. By the Four Point Lemma, we can find a projective map taking A_1, A_2, A_3 respectively to E_1, E_2, E_3. The conic F then maps to an irreducible conic through these three points, so having the form $hxy + fyz + gzx = 0$ for some scalars f, g, h, necessarily non-zero (else the conic reduces). The claim follows by observing that the scaling $x = fX$, $y = gY$, $z = hZ$ leaves E_1, E_2, E_3 fixed and maps this conic to $xy + yz + zx$.

Step 2: Consider now a cubic of conic plus chord type, reducing to an irreducible conic F, and a line L, meeting F in two distinct points A_1, A_2. And let A_3 be any other point on F distinct from A_1, A_1. By Step 1 there is a projective map taking A_1, A_2, A_3 respectively to E_1, E_2, E_3, and F to the conic $xy + yz + zx$. In particular, the chord L will be mapped to the line $z = 0$ through E_1, E_2. Thus the cubic is projectively equivalent to $z(xy + yz + zx)$. And since the projective maps form a group, any two cubics of conic plus chord type will be projectively equivalent.

Step 3: Finally, consider a cubic of conic plus tangent type, reducing to an irreducible conic F, and a line L, tangent to F at a points A_1. And let A_2, A_3 be any other points on F distinct from A_1 and each other. Then, again by Step 1 there is a projective map taking A_1, A_2, A_3 respectively to E_1, E_2, E_3, and F to the conic $xy + yz + zx$. In particular, the tangent L will be mapped to the tangent $y + z$ to $xy + yz + zx$ at E_1. Thus the cubic is projectively equivalent to $(y + z)(xy + yz + zx)$. As in Step 2, we conclude that any two cubics of conic plus tangent type are projectively equivalent. □

Lemma 15.7 *Any two cubics in* $\mathbb{PC}^2$ *which reduce to three lines, and are of the same geometric type, are projectively equivalent.*

Proof We will illustrate the proof by considering cubics which reduce to three concurrent lines. (The other three types are handled in Exercises 15.4.5, 15.4.6, and 15.4.7 below by small modifications of the argument.) It suffices to show that any such cubic F is projectively equivalent to the curve $(y - z)(z - x)(x - y)$. Let P be the point of intersection of the three lines. And let Q_1, Q_1, Q_1 be three points distinct from P, one on each of the three lines. By the Four Point Lemma, there is a projective equivalence taking P, Q_1, Q_1, Q_1 respectively to the points U, E_1, E_2, E_3 defined above, and taking F to a curve G of the same geometric type. The lines are then the lines joining E_1, E_2, E_3 to U, i.e. $y = z$, $z = x$, $x = y$. It follows from Lemma 10.1 that $y - z$, $z - x$, $x - y$ are all components of G, and hence that $G = (y - z)(z - x)(x - y)$, up to scalar multiples. □

Exercises

15.4.1 $F = x^3 + 5x^2y + 3xy^2 - 9y^3 + x^2z + 4xyz + 3y^2z$ defines a cubic curve in $P\mathbb{C}^2$. Determine the singular points of F, and deduce that F is reducible. Find the components of F.

15.4.2 What are the geometric types of the cubic curves F, G in $P\mathbb{C}^2$ defined by $F = x(x - z)(x + z)$ and $G = y(y - z)(y + z)$? A cubic curve H is defined by $H = \alpha F + \beta G$, where α, β are scalars, at least one of which is non-zero. Show that H is singular if and only if $\alpha = 0$, $\beta = 0$, $\alpha + \beta = 0$ or $\alpha - \beta = 0$. Assume that $\alpha \neq 0$, $\beta \neq 0$. Show that when $\alpha + \beta = 0$, or $\alpha - \beta = 0$, H is of chord plus conic type, and find the equations of the chords and the conics.

15.4.3 Describe the four possible ways in which a cubic curve in $P\mathbb{C}^2$ can reduce to three lines. Let F be a cubic curve in $P\mathbb{R}^2$, for which the corresponding curve in $P\mathbb{C}^2$ reduces to three lines. Describe the possible ways in which F can reduce.

15.4.4 Let F be a cubic curve in the real projective plane $P\mathbb{R}^2$ which reduces to a line L and a (real or imaginary) line-pair C meeting L only at its vertex V. Show that when C is real, F is projectively equivalent to $y(x^2 - y^2)$, and that when C is imaginary, F is projectively equivalent to $y(x^2 + y^2)$. (Apply the Four Point Lemma to two points, and then make explicit changes of coordinates.)

15.4.5 Show that any two cubics in $P\mathbb{C}^2$ of triangle type are projectively equivalent. (Follow the line of reasoning in Lemma 15.7.)

15.4.6 Show that any two cubics in $P\mathbb{C}^2$ which reduce to two distinct lines, one of multiplicity 2, are projectively equivalent.

15.4.7 Show that any two cubics in $P\mathbb{C}^2$ which reduce to a single line of multiplicity 3 are projectively equivalent.

16

Linear Systems

A long established philosophy in mathematics is that when you have an interesting set of objects you wish to study you should consider them as the 'points' of some 'space', and then think in terms of the geometry of that space. The objective of this chapter is to put flesh on this idea, in the case when the interesting objects are projective curves of degree d in $\mathbb{PK}^2$. This breaks down into two steps. The first is to extend affine space $\mathbb{K}^n$ to projective space $\mathbb{PK}^n$ by adding points at infinity, generalizing the construction of Chapter 9 given in the case $n = 2$. And the second step is to show that the curves of degree d in $\mathbb{PK}^2$ form a projective space $\mathbb{PK}^D$, where $D = \frac{1}{2}d(d+3)$. That enables one to use the machinery of linear algebra to discuss questions about curves. Of particular importance to us will be 'pencils' of curves of degree d, corresponding to lines in $\mathbb{PK}^D$.

16.1 Projective Spaces of Curves

We proceed by analogy with the construction of the projective plane in Section 9.1. We define *projective n-space* $\mathbb{PK}^n$ to be the set of lines through the origin $O = (0,\ldots,0)$ in $\mathbb{K}^{n+1}$. These lines are the *points* of $\mathbb{PK}^n$. Given a non-zero vector $X = (x_1,\ldots,x_{n+1})$ in $\mathbb{K}^{n+1}$, there is a unique line through X, O determining a unique point in $\mathbb{PK}^n$ denoted $(x_1 : \ldots : x_{n+1})$. The vector X is a *representative* for the point, and its entries $x_1, \ldots, x_{n+1}$ are *homogeneous coordinates* for that point. They are not unique, for any scalar $\lambda \neq 0$ the vector $\lambda X = (\lambda x_1,\ldots,\lambda x_{n+1})$ represents the same point in $\mathbb{PK}^n$.

Example 16.1 The very simplest case of the above construction is the case $n = 1$ of the *projective line* $\mathbb{PK}^1$. Points are lines through the origin in $\mathbb{K}^2$, and are identified with ratios $(x : y)$; explicitly, the line $ax+by = 0$

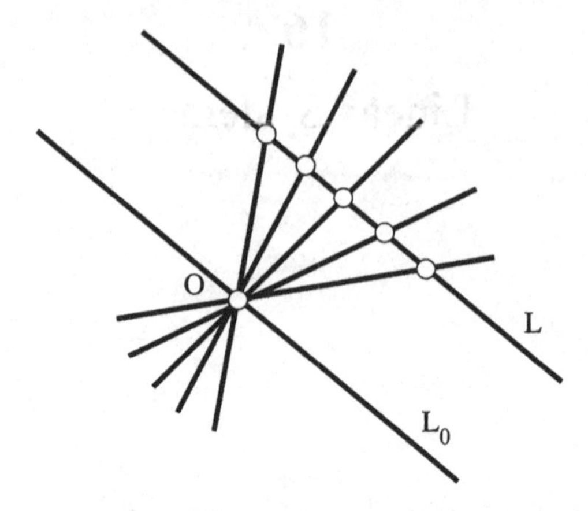

Fig. 16.1. The real projective line

through O corresponds to the ratio $(-b : a)$. The concept of an 'affine view' can be introduced exactly as in the case of the projective plane. We start from an arbitrary affine line L in $\mathbb{K}^2$ not through the origin O, and denote by L_0 the parallel line through O. We then have a natural one-to-one correspondence between points on L and lines through O in $\mathbb{K}^2$ distinct from L_0. (The situation is illustrated in Figure 16.1 for the case of the real field.) The line L_0 defines the *point at infinity* associated to L. Thus we can think of $\mathrm{P}\mathbb{K}^1$ as the affine line L, plus a single point at infinity. For practical purposes, one can take L to be the line $y = 1$. With that choice, points x on the affine line are identified with points $(x : 1)$ on the projective line, and the point at infinity is the ratio $\infty = (1 : 0)$.

The projective spaces of key importance to us arise from the following proposition.

Lemma 16.1 *The curves of degree d in $\mathrm{P}\mathbb{K}^2$ form a projective space $\mathrm{P}\mathbb{K}^D$, where $D = \frac{1}{2}d(d+3)$.*

Proof Let $m_1, \dots, m_{D+1}$ be the monomials of degree d in x, y, z, i.e. of the form $x^i y^j z^k$ where $i + j + k = d$. An algebraic curve of degree d in $\mathrm{P}\mathbb{K}^2$ has the form $\sum a_k m_k$ with the a_k in $\mathbb{K}$, and at least one non-zero; we identify this curve with the point $(a_1 : \dots : a_{D+1})$ in $\mathrm{P}\mathbb{K}^D$. That yields a one-to-one correspondence between the curves of degree d and the points of $\mathrm{P}\mathbb{K}^D$. It remains to compute D. The monomials in x, y, z of degree d

are those in the following array

$$\begin{array}{ccccccc} & & & z^d & & & \\ & & xz^{d-1} & & yz^{d-1} & & \\ & x^2z^{d-2} & & xyz^{d-2} & & y^2z^{d-2} & \\ & \vdots & & \vdots & & \vdots & \\ x^d & x^{d-1}y & \cdots & \cdots & \cdots & xy^{d-1} & y^d. \end{array}$$

There are k monomials in the kth row, and $(d+1)$ rows in all. Thus the total number of monomials of degree d is

$$D+1 = 1+2+\cdots+(d+1) = \frac{1}{2}(d+1)(d+2)$$

yielding the required result. □

Example 16.2 A line in $\mathbb{PK}^2$ has the form $ax+by+cz$, identified with a point $(a:b:c)$ in $\mathbb{PK}^2$, so lines in $\mathbb{PK}^2$ form another projective plane $\mathbb{PK}^2$, called the *dual* plane.

Example 16.3 A conic $ax^2+by^2+cz^2+2dxy+2eyz+2fzx$ in $\mathbb{PK}^2$ is identified with a point $(a:b:c:d:e:f)$ in $\mathbb{PK}^5$, so conics in $\mathbb{PK}^2$ form a projective space $\mathbb{PK}^5$.

16.2 Pencils of Curves

Given two distinct points $A = (a_1 : \ldots : a_{n+1})$, $B = (b_1 : \ldots : b_{n+1})$ in a projective space $\mathbb{PK}^n$, we define the *parametrized line* through A, B to be the set of all points $\lambda A + \mu B$ with at least one of λ, μ non-zero. (In the case $n = 2$, this agrees with the definition given in Chapter 9.) We are going to be interested in the case of the projective space $\mathbb{PK}^D$ of curves of degree d in $\mathbb{PK}^2$. In that case, a parametrized line has the form $\lambda F + \mu G$ where F, G are curves in $\mathbb{PK}^2$ of degree d, and is called a *pencil* of curves of degree d.

Example 16.4 The set of all lines in $\mathbb{PK}^2$ through a fixed point $P = (\alpha : \beta : \gamma)$ is a pencil. Indeed it comprises all lines $ax+by+cz$ with $a\alpha + b\beta + c\gamma = 0$, which is a single linear condition on a, b, c defining a line in the dual plane. The idea is illustrated in Figure 16.2, each line

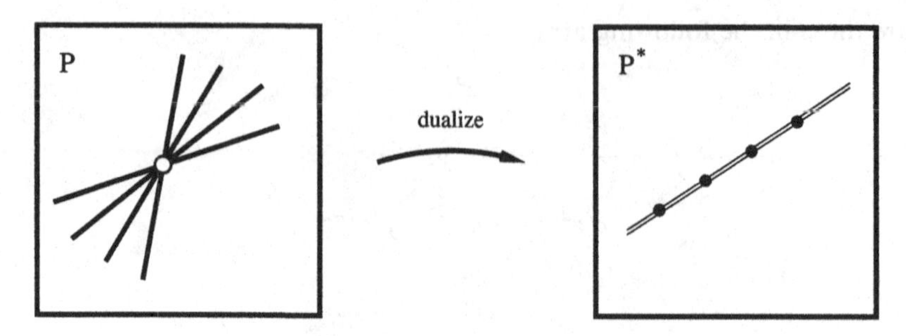

Fig. 16.2. Pencil of lines through a point

through P corresponds to a point on the line in the dual plane, and conversely.

Example 16.5 Recall that the general circle in $\mathbb{R}^2$ with centre (a, b) and radius $r > 0$ has equation $(x-a)^2 + (y-b)^2 = r^2$. Consider the centre as fixed, and the radius as variable. The corresponding projective curve in $P\mathbb{R}^2$ is $(x - az)^2 + (y - bz)^2 = r^2z^2$. Clearly, all such curves occur within the pencil of real conics $\lambda F + \mu G$, where $F = (x - az)^2 + (y - bz)^2$ (an imaginary line-pair) $G = z^2$ (a repeated line). The affine type of the curve $\lambda F + \mu G$ in its affine view $z = 1$ is readily established; it is a real ellipse when $\lambda/\mu < 0$, an imaginary line-pair when $\lambda/\mu = 0$, and an imaginary ellipse when $\lambda/\mu > 0$. Thus all the circles with fixed centre (a, b) comprise *part* of the resulting projective pencil.

Example 16.6 The set $\mathscr{L}$ of all conics in $P\mathbb{K}^2$ which pass through $A = (1:0:0)$, $B = (0:1:0)$, $C = (0:0:1)$, $D = (1:1:1)$ is a pencil. Let $F = ax^2 + by^2 + cz^2 + 2dxy + 2eyz + 2fzx$ be a general conic. The conditions for F to pass through A, B, C, D respectively are that $a = 0$, $b = 0$, $c = 0$, $a + b + c + 2d + 2e + 2f = 0$. Thus $F = dxy + eyz + fzx$ with $e + d + f = 0$, i.e. $F = dxy + eyz - (d + e)zx = dx(y - z) + ez(y - x)$ where e, d are arbitrary, but not both zero. Thus the conics F through A, B, C, D form a pencil, spanned by the line-pairs $x(y - z)$, $z(y - x)$.

Example 16.7 Recall that a Steiner cubic is a cubic curve of the form $\mu(x^3 + y^3 + z^3) + 3\lambda xyz$ in $P\mathbb{C}^2$ with at least one of λ, μ non-zero. In Example 13.6, we saw that the flexes of the Steiner cubics comprise the nine point configuration of the points P_{ij} with $0 \leq i, j \leq 2$. (The coordinates of these points were written out explicitly in that example.)

And in Lemma 13.4 we saw that any cubic through these nine points is necessarily a Steiner cubic. Thus *the cubics which pass through the nine point Steiner configuration form a pencil*, spanned by $F = xyz$ and $G = x^3 + y^3 + z^3$. In Chapter 17 we will see that this is a special case of an important general result, namely the Theorem of the Nine Associated Points.

Lemma 16.2 *Any two distinct curves F', G' in a pencil $\lambda F + \mu G$ have the same points of intersection as F, G.*

Proof Write $F' = \alpha F + \beta G$, $G' = \gamma F + \delta G$ for some scalars α, β, γ, δ with $\alpha\delta - \beta\gamma \neq 0$. Then, for any point P in the plane we have

$$\begin{cases} F'(P) &= \alpha F(P) + \beta G(P) \\ G'(P) &= \gamma F(P) + \delta G(P) \end{cases}$$

so $F(P) = G(P) = 0$ if and only if $F'(P) = G'(P) = 0$ i.e. F', G' have the same points of intersection P as F, G. □

This result may help to determine the intersection of F, G, especially if F', G' can be *chosen* to be reducible.

Example 16.8 Consider two distinct conics F, G in $P\mathbb{C}^2$. Then $\lambda F + \mu G$ is a complex conic, so can be identified with a symmetric 3×3 matrix of coefficients, and is reducible if and only if it is singular, i.e. if and only if $\det(\lambda F + \mu G) = 0$. That is a binary cubic equation in the ratio $(\lambda : \mu)$ with three roots, counted properly. Thus (in general) we expect three line-pairs in the pencil, and the intersection of any two of them will give the intersections of F, G. Assume that F, G have no common component. In that case, the two line-pairs have no common line, so intersect in ≤ 4 points. Thus we deduce that *two conics in $P\mathbb{C}^2$ with no common line component intersect in ≤ 4 points*; of course we already know this from Bézout's Theorem, but it is illuminating to see a completely different approach to the same statement. Moreover, this makes it clear why we expect *three* line-pairs in a pencil of conics. Given four points A, B, C, D there are three 'obvious' line-pairs passing through them all: the line-pair comprising the lines AB, CD, the line-pair comprising the lines AD, BC, and the line-pair comprising the lines AC, BD. (Figure 16.3.)

Example 16.9 Let $f = x^2 + y^2 - a^2$, $g = x^2 + y^2 - b^2$ be the circles in $\mathbb{R}^2$ centred at the origin with $0 < a < b$, giving rise to the projective conics

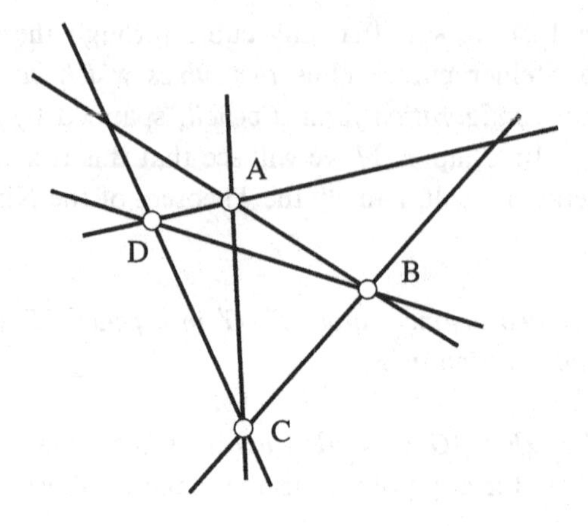

Fig. 16.3. Three line-pairs in a pencil of conics

$F = x^2 + y^2 - a^2z^2$, $G = x^2 + y^2 - b^2z^2$ in $P\mathbb{C}^2$. We know already that F, G intersect at the circular points at infinity I, J, in fact these are the *only* intersections. Although that is very easily demonstrated by direct calculation, it is illuminating to see how it fits into the above pattern. The conics F, G determine a pencil $\lambda F + \mu G$; the symmetric matrix associated to the conic $\lambda F + \mu G$ is

$$\begin{pmatrix} \lambda+\mu & 0 & 0 \\ 0 & \lambda+\mu & 0 \\ 0 & 0 & -(\lambda a^2 + \mu b^2) \end{pmatrix}.$$

The determinant is the binary cubic $-(\lambda+\mu)^2(\lambda a^2+\mu b^2)$ having a repeated zero $(-1 : 1)$ (corresponding to the repeated line z^2) and a simple zero $(-b^2 : a^2)$ (corresponding to the line-pair x^2+y^2). These reducible conics intersect only at I, J so by Lemma 16.2 the same is true of F, G. Note that in this example the degeneracy in the binary cubic (two of the roots have coalesced) gives rise to a degeneracy in the line-pairs (one has reduced to a repeated line).

Exercises

16.2.1 Show that the set $\mathscr{L}$ of all conics F in $P\mathbb{C}^2$ which pass through the points $P = (1 : 0 : 0)$, $Q = (0 : 1 : 0)$, $R = (0 : 0 : 1)$ and are tangent at R to the line $y = x$ form a pencil.

16.2.2 Let $\mathscr{L}$ be the set of all conics in $P\mathbb{R}^2$ which pass through the points $A = (1 : 0 : 1)$, $B = (0 : 1 : 0)$, $C = (0 : 0 : 1)$ and are tangent at B to the line $z = 0$. Show that $\mathcal{L}$ is a pencil, and that in the affine plane defined by $z = 1$ it is the family of parabolas $y = \lambda x(x - 1)$, with λ a ratio.

16.2.3 Write down the equation of a general conic C in $\mathbb{PC}^2$ which passes through the point $O = (0 : 0 : 1)$ and the circular points at infinity I, J. Find the condition for such a conic to be tangent to the line $y = 0$ at O. Deduce that the set $\mathcal{L}$ of all conics through O, I, J which are tangent to $y = 0$ at O is a pencil.

16.2.4 Let $\mathcal{L}$ be the set of all conics in $\mathbb{PC}^2$ which pass through the point $O = (0 : 0 : 1)$, and the circular points at infinity I, J, are tangent at I to the line IO, and are tangent at J to the line JO. Show that $\mathcal{L}$ is a pencil of conics, and that any two conics in $\mathcal{L}$ intersect only at the points I and J. Describe the conics in $\mathbb{R}^2$ having at least one real point, for which the corresponding conic in $\mathbb{PC}^2$ is in $\mathcal{L}$.

16.2.5 Find the reducible conics in the pencil $\lambda C + \mu D$ where C, D are the conics in $\mathbb{PC}^2$ defined by $C = 2(y^2 + 2xy + xz + yz + z^2)$, $D = 2(x^2 - xz - yz - z^2)$. Hence find the points of intersection of C, D.

16.2.6 Conics C, D in $\mathbb{PC}^2$ are defined by $C = 2(xy + yz - 2zx)$ and $D = 2(xy + 2yz - 3zx)$. Write down the symmetric 3×3 matrix corresponding to the conic $E = \lambda C + \mu D$. Show that E is reducible if and only if $\lambda + \mu = 0$, $2\lambda + 3\mu = 0$ or $\lambda + 2\mu = 0$. In each of these cases, find the equations of the lines comprising the line-pair E. Find the intersections of C, D.

16.3 Solving Quartic Equations

Here is an application of pencils of conics to algebra. The general theory of polynomial equations in a single variable predicts that the problem of solving a *quartic* equation $q(x) = 0$ over $\mathbb{C}$ can be reduced to that of solving an auxiliary *cubic* equation. There is a geometric viewpoint of this problem, which dates back to classical Greek mathematics.

Example 16.10 Menaechmus sought a geometric construction of the real number $x = 2^{\frac{1}{3}}$, which satisfies the quartic equation $x^4 - 2x = 0$. Setting $y = x^2$ we see that the algebraic problem reduces to the geometric one of finding the intersections of the parabolas $f = y - x^2$ and $g = y^2 - 2x$. Thus the key was to find a geometric construction for a parabola.

This construction is quite general. Note first that we can suppose the coefficient of x^3 in $q(x)$ is zero, by replacing x by $x+\alpha$ for an appropriately chosen α. Write the resulting equation as $q(x) = x^4+rx^2+sx+t = 0$, and let the roots be a_1, a_2, a_3, a_4. Recall at this point that the roots satisfy the *Newton relations*.

$$\begin{cases} 0 &= a_1 + a_2 + a_3 + a_4 \\ r &= a_1a_2 + a_2a_3 + a_3a_4 + a_4a_1 + a_1a_3 + a_2a_4 \\ -s &= a_1a_2a_3 + a_1a_2a_4 + a_1a_3a_4 + a_2a_3a_4 \\ t &= a_1a_2a_3a_4. \end{cases}$$

Set $y = x^2$ and observe that the problem is the geometric one of finding the intersections $p_k = (a_k, a_k^2)$ of the parabolas $f = y - x^2$ and $g = y^2+ry+sx+t$. Projectively, we seek the intersections $P_k = (a_k : a_k^2 : 1)$ of the conics $F = yz-x^2$, $G = y^2+ryz+szx+tz^2$. (Note that they do not meet at infinity.) In principle, the pencil of conics spanned by F, G contains three line-pairs. We can determine them explicitly. Write L_{ij} for the line joining P_i, P_j. By computation, L_{ij} has equation $y-(a_i+a_j)x+a_ia_jz = 0$ and the three-line-pairs are $L_{12}L_{34}$, $L_{13}L_{24}$, $L_{14}L_{23}$. We should be able to write each of these in the form $F - \lambda G$ for some ratio λ. Indeed, a minor computation using the Newton relations shows that

$$\begin{cases} L_{12}L_{34} &= F-(a_1+a_2)(a_3+a_4)G \\ L_{13}L_{24} &= F-(a_1+a_3)(a_2+a_4)G \\ L_{14}L_{23} &= F-(a_1+a_4)(a_2+a_3)G. \end{cases}$$

The *auxiliary cubic* is defined to be the cubic polynomial in one variable having the roots $-(a_1+a_2)(a_3+a_4)$, $-(a_1+a_3)(a_2+a_4)$, $-(a_1+a_4)(a_2+a_3)$. In algebra texts, the auxiliary cubic is computed via elementary symmetric functions; the geometry shows that it is given by $\det(F-\lambda G) = 0$, where the conics F, G are identified with 3×3 symmetric matrices.

16.4 Subspaces of Projective Spaces

To any subspace $U \subseteq \mathbb{K}^{n+1}$ of dimension $(m+1)$ we can associate the set $PU \subseteq P\mathbb{K}^n$ comprising all lines through O in $\mathbb{K}^{n+1}$ which lie in U. The set PU is a (projective) *subspace* of $P\mathbb{K}^n$ of (projective) *dimension* m, and (projective) *codimension* $(n-m)$. The cases of this construction we will meet most frequently are gathered together in Table 16.1. The first column gives the dimension of U, the second column gives the name usually associated to PU, the third gives its projective dimension (one less than the affine dimension) and the fourth gives the projective codimension.

Table 16.1. *Projective subspaces*

dim U	name	proj dim	proj codim
0	empty set	-1	$n+1$
1	point	0	n
2	line	1	$n-1$
3	plane	2	$n-2$
n	hyperplane	$n-1$	1

The following remarks are immediate from standard results in linear algebra. If PU has codimension c then U can be defined by c linearly independent linear conditions on the homogeneous coordinates $x_1, \ldots, x_{n+1}$; more generally, if U is defined by d (not necessarily independent) linear conditions on $x_1, \ldots, x_{n+1}$ then $c \leq d$. Here is a useful fact we will appeal to several times.

Lemma 16.3 *Let $PU_1, \ldots, PU_s$ be projective subspaces of* $\mathrm{P}\mathbb{K}^n$ *of respective codimensions $c_1, \ldots, c_s$: then the intersection $PU_1 \cap \ldots \cap PU_s$ has codimension $\leq c_1 + \cdots + c_s$.* (Subadditivity of the codimension.)

Proof Observe that the intersection is defined by $c_1 + \cdots + c_s$ linear conditions in total, and apply the above remarks. □

Example 16.11 The simplest case of the above is when $n = 2$. A plane U through the origin in $\mathbb{K}^3$ is a linear subspace, given by a single linear condition $ax + by + cz = 0$ on $\mathbb{K}^3$, and defining a line in the projective plane $\mathrm{P}\mathbb{K}^2$ of projective codimension 1. Now take two lines in $\mathrm{P}\mathbb{K}^2$, each defined by a single linear condition on x, y, z. Then (by subadditivity) the codimension of the intersection is ≤ 2. If the lines are distinct (so the linear conditions are independent), they intersect in a single point of codimension 2, but if the lines coincide (so the linear conditions are dependent), they intersect in a line, of codimension 1.

The concept of a 'projective map' can be extended to general projective spaces. Let L be an invertible linear mapping of $\mathbb{K}^{n+1}$. Then the image under L of any line through the origin is another. Thus L defines a (bijective) mapping $\tilde{L}$ of $\mathrm{P}\mathbb{K}^n$, called a *projective* map of $\mathrm{P}\mathbb{K}^n$. Subspaces of $\mathrm{P}\mathbb{K}^n$ are mapped by projective maps to subspaces of the same dimension. Projective mappings arise very naturally in the context of the projective space $\mathrm{P}\mathbb{K}^D$ of projective curves of degree d in $\mathrm{P}\mathbb{K}^2$. Any projective map of $\mathbb{PK}^2$ is induced by an invertible linear mapping Φ of $\mathbb{K}^3$. That induces a linear mapping L_Φ of $\mathbb{K}^{D+1}$ defined by $L_\Phi = F \circ \Phi$; moreover, L_Φ is invertible, with inverse $L_{\Phi^{-1}}$. Thus *any projective mapping of* $\mathbb{PK}^2$ *induces a projective mapping of the space* $\mathbb{PK}^D$ *of curves of degree d.* This simple remark is one of the keys to the next section, it means (for instance) that we can apply the Four Point Lemma in $\mathbb{PK}^2$ without changing the dimensions of projective subspaces.

Exercises

16.4.1 Show that n hyperplanes in $\mathbb{PK}^n$ have at least one common point.

16.5 Linear Systems of Curves

Projective subspaces of the space $\mathbb{PK}^D$ of projective curves of degree d in $\mathbb{PK}^2$ are called *linear systems* of curves. We will denote them by upper case calligraphic letters, such as $\mathscr{L}, \mathscr{M}, \mathscr{N}, \ldots$ The simplest possible linear systems are those of dimension 1, arising from planes through the origin in $\mathbb{K}^{D+1}$, comprising all elements of the form $sF + tG$ for fixed F, G, thus linear systems of dimension 1 are simply the pencils studied at length in Section 16.2. In this section, we will look at linear systems of curves of arbitrary dimension which arise naturally in geometric situations.

Lemma 16.4 *The projective curves F of degree d which pass through a given point P in* $\mathbb{PK}^2$ *with multiplicity* $\geq m$ *on F form a linear system* $\mathscr{L}$ *in* $\mathbb{PK}^D$ *of codimension* $\frac{1}{2}m(m+1)$. *In particular, the curves of degree d which pass through P form a hyperplane in* $\mathbb{PK}^D$.

Proof In view of the remarks in the previous section, we can apply the Four Point Lemma and assume that $P = (0 : 0 : 1)$. Write $F = F_0(x, y)z^d + F_1(x, y)z^{d-1} + \cdots + F_d(x, y)$ where each $F_k(x, y)$ is a form of degree k in x, y. The condition for F to have multiplicity $\geq m$ at P is that $F_0(x, y) \equiv 0, \ldots, F_{m-1}(x, y) \equiv 0$, i.e. that the coefficients of all monomials $x^i y^j z^k$ in F with $i + j < m$ vanish. The result follows from the fact that there are $1 + 2 + \cdots + m = \frac{1}{2}m(m+1)$ such monomials. The particular case arises on taking $m = 1$. □

Lemma 16.5 *The projective curves F of degree d which pass through given points* $P_1, \ldots, P_s$ *with respective multiplicities* $\geq m_1, \ldots, \geq m_s$ *form a linear system* $\mathscr{L}$ *of codimension* $\leq \sum \frac{1}{2}m_k(m_k+1)$. *In particular, the projective* *curves of degree d which pass through given points $P_1, \ldots, P_s$ form a linear system of codimension $\leq s$.*

Proof Let $\mathscr{L}_k$ be the projective subspace of all curves of degree d which pass through P_k with multiplicity $\geq m_k$. By Lemma 16.4, the linear system $\mathscr{L}_k$ has codimension $\frac{1}{2}m(m+1)$. Clearly, $\mathscr{L} = \mathscr{L}_1 \cap \ldots \cap \mathscr{L}_s$ and subadditivity of the codimension yields cod $\mathscr{L} \leq \sum \text{cod } \mathscr{L}_k = \sum \frac{1}{2}m_k(m_k+1)$. The particular case arises on taking $m_1 = 1, \ldots, m_s = 1$. □

Here are some useful applications of these ideas, illustrating how one can obtain geometric information about projective curves simply on the basis of linear algebra.

Example 16.12 The linear system $\mathscr{L}$ of conics in $\mathbb{PK}^2$ which pass through *any* four points A, B, C, D in general position is a pencil. Indeed, by the Four Point Lemma, we can assume that $A = (1 : 0 : 0)$, $B = (0 : 1 : 0)$, $C = (0 : 0 : 1)$, $D = (1 : 1 : 1)$, and then the result is immediate from the calculation in Example 16.6.

Lemma 16.6 *Let $\mathscr{L}$ be a linear system of curves of degree d having dimension s, and let $P_1, \ldots, P_s$ be points in $\mathbb{PK}^2$. Then there exists at least one curve in $\mathscr{L}$ which passes through all of $P_1, \ldots, P_s$.*

Proof Let $\mathscr{L}'$ be the linear system of curves in $\mathscr{L}$ passing through $P_1, \ldots, P_s$. Then $\mathscr{L}'$ is defined by the $(D - s)$ linear conditions defining $\mathscr{L}$, and the s linear conditions imposed by the curve passing through $P_1, \ldots, P_s$, so in total $(D - s) + s = D$ conditions. Thus $\mathscr{L}'$ has codimension $\leq D$, hence dimension ≥ 0. In particular, $\mathscr{L}'$ contains at least one curve. □

Example 16.13 A special case of Lemma 16.6 is that *in a pencil of curves of degree d there is at least one curve passing through any given point P*. Good illustrations are the pencil of lines through a given point P, or the pencil of circles with given centre P.

Example 16.14 Another special case of Lemma 16.6 arises when $\mathscr{L}$ is taken to be the linear system of *all* curves of degree d, so having dimension D. We then see that there is always at least one curve of degree d through D points in $\mathbb{PK}^2$. Thus when $d = 1$, this just says there is at least one line through two points; when $d = 2$, there is at least one conic through five points; when $d = 3$, there is at least one cubic through nine points. And so on.

Example 16.15 Here is another way of doing Example 16.12, with a slightly weaker hypothesis. The linear system $\mathscr{L}$ of conics in $\mathbb{PK}^2$ which pass through *any* four distinct points A, B, C, D is a pencil, *provided the points are not collinear*, a hypothesis which certainly holds when the points are in general position. The linear system $\mathscr{L}$ has codimension ≤ 4 by Lemma 16.4, hence dimension ≥ 1. Suppose $\mathscr{L}$ has dimension ≥ 2. We derive a contradiction as follows. Choose E to be any point on the line M joining C, D distinct from A, B, C, D. Then M meets any conic through A, B, C, D in three distinct points, so must be a component of that conic. The other component must be the line L joining A, B and distinct from M, else the four points lie on a line. Now choose F to be any point which lies neither on L nor on M. By Lemma 16.6, there exists a conic in $\mathscr{L}$ through E, F, i.e. a conic through A, B, C, D, E and F, yielding the desired contradiction. In the exceptional case when A, B, C, D lie on a single line N, any conic through A, B, C, D would have N as a component, and the linear system of all conics through A, B, C, D would comprise all line-pairs with N as a component, so would be 2-dimensional.

Exercises

16.5.1 Determine the unique conic F in $\mathbb{PC}^2$ which passes through the five points $(1 : 0 : 0)$, $(0 : 1 : 0)$, $(0 : 0 : 1)$, $(1 : 1 : 1)$ and $(1 : 2 : 3)$. Verify that F is irreducible.

16.5.2 Let P be a point in $\mathbb{PK}^2$, and let L be a line through P. Show that the set $\mathscr{L}$ of all conics which pass through P, and are tangent there to L, is a linear system of codimension 2.

16.5.3 Let P be a point in $\mathbb{PK}^2$, and let L be a line through P. Show that the set $\mathscr{L}$ of all curves of degree $d \geq 2$ which pass through P, and are tangent there to L, is a linear system of codimension 2. (The previous exercise is the case $d = 2$ of this result.)

16.5.4 Let P be a point in $\mathbb{PK}^2$, let L be a line through P, and let $n \geq 2$ be an integer. Show that for $d \geq n$ the set $\mathscr{L}$ of all curves F of degree d which pass through P, and for which $I(P, F, L) \geq n$ forms a linear system of codimension n. (The previous exercise is the case $n = 2$ of this result.)

16.5.5 In $\mathbb{PC}^2$ three points I, J, O are defined by $I = (1 : i : 0)$, $J = (1 : -i : 0)$, $O = (0 : 0 : 1)$. Let $\mathscr{L}$ be the set of all conics in $\mathbb{PC}^2$ which pass through I and J, are tangent at I to the line IO, and are tangent at J to the line JO. Show that $\mathscr{L}$ is a pencil of conics, and that any two conics in $\mathscr{L}$ intersect only at the points I and J.

16.5.6 Let $P_1, \dots, P_n$ be distinct points in $\mathbb{PK}^2$. Show that if $d \geq n-1$ then the linear system of all curves of degree d through $P_1, \dots, P_n$ has codimension n.

16.5.7 Let $D = \frac{1}{2}d(d+3)$, and let $1 \leq n \leq D$. Show, by induction on n, that there exist n points $P_1, \dots, P_n$ in $\mathbb{PC}^2$ such that the linear system $\mathscr{L}$ of all projective curves of degree d passing through $P_1, \dots, P_n$ has codimension precisely n. (Compare with Lemma 16.5.)

16.6 Dual Curves

This is as good a place as any to say something about the interesting and illuminating concept of 'duality'. It will not play any formal role in the rest of this text, but anyone wishing to pursue the geometry of curves further will find the idea indispensable. It turns on the fact, mentioned at the beginning of this chapter, that the lines in $\mathbb{PK}^2$ form a projective plane, called the *dual* plane. For the sake of simplicity, we will write $\mathbb{P}$ for the projective plane, and $\mathbb{P}^\star$ for the dual plane. Write x, y, z for homogeneous coordinates in $\mathbb{P}$, and X, Y, Z for homogeneous coordinates in $\mathbb{P}^\star$. Recall that for fixed X, Y, Z the line $Xx+Yy+Zz$ in $\mathbb{P}$ corresponds to the point $(X : Y : Z)$ in $\mathbb{P}^\star$, and conversely for fixed x, y, z the line $Xx+Yy+Zz$ in $\mathbb{P}^\star$ defines a pencil of lines in $\mathbb{P}$, through the point $(x : y : z)$ in $\mathbb{P}$. In this way we identify $(\mathbb{P}^\star)^\star$ with $\mathbb{P}$. We denote the line in $\mathbb{P}^\star$ dual to a point P by $P^\star$, and the point in $\mathbb{P}$ dual to a line L in $\mathbb{P}^\star$ by $L^\star$. Thus $(P^\star)^\star = P$, $(L^\star)^\star = L$.

The main feature of this correspondence is that it preserves 'incidence' in the sense that the point P lies on the line L in $\mathbb{P}$ if and only if the line $P^\star$ passes through the point $L^\star$ in $\mathbb{P}^\star$. This is the origin of the historically important Duality Principle, namely that results about the incidence of points and lines in $\mathbb{P}$ translate into results about the incidence of lines and points in $\mathbb{P}^\star$, when we interchange the words 'point' and 'line', and the phrases 'the point P lies on the line L' and 'the line $P^\star$ passes through the point $L^\star$'. For instance, the statement that a set of lines L_i pass through the point P in $\mathbb{P}$ translates to the statement that the corresponding set of

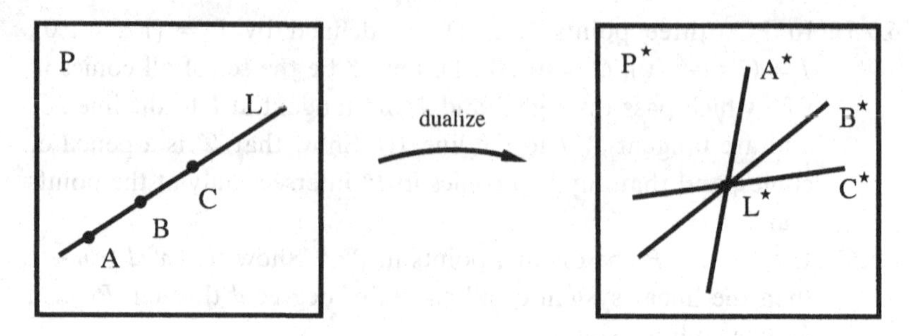

Fig. 16.4. Dual concepts of collinearity and concurrency

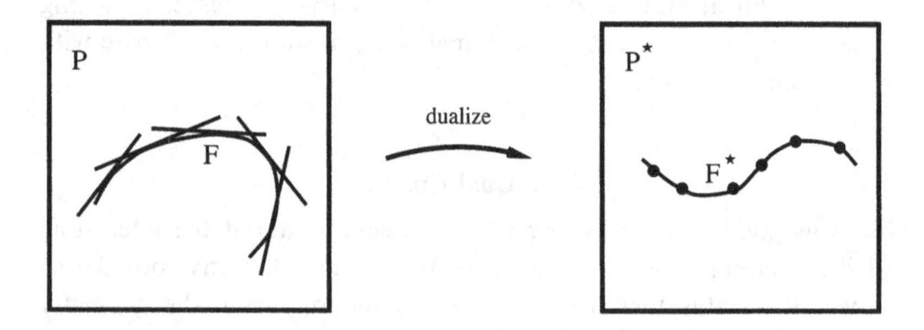

Fig. 16.5. The concept of the dual curve

points $L_i^\star$ lie on the line $P^\star$ in $\mathbb{P}^\star$: thus the concepts of 'concurrency' and 'collinearity' are dual to each other in the projective plane. (Figure 16.4.)

One can go further down this road. At each point P on a curve F in the projective plane $\mathbb{P}$, we have tangents (normally just one, but more at a singularity) which correspond to points in the dual plane $\mathbb{P}^\star$. The intuitive idea (Figure 16.5) is that as P traces out the curve F so these points will trace out a dual curve $F^\star$.

A formal treatment of this idea requires more algebra than we have to hand. We will instead illustrate it in a special case. Let $Q(x, y, z)$ be a *non-singular* conic. Then at any point $P = (x : y : z)$ there is a tangent line $Q_x(P)x + Q_y(P)y + Q_z(P)z = 0$ yielding a point $(X : Y : Z)$ in $\mathbb{P}^\star$, where $X = Q_x(P)$, $Y = Q_y(P)$, $Z = Q_z(P)$. Each of the partial derivatives is linear in x, y, z since Q is quadratic in these variables. The idea is to solve these linear equations for x, y, z in terms of X, Y, Z, then substitution in $Xx + Yy + Zz = 0$ yields (in principle) an equation of degree 2 in X, Y, Z. Thus we expect the 'dual' of $Q(x, y, z)$ to be another conic $Q^\star(X, Y, Z)$.

More concretely, consider a non-singular conic Q given by a formula

$$Q = ax^2 + by^2 + cz^2 + 2hxy + 2fyz + 2gzx.$$

Then the condition for $(X : Y : Z)$ to represent a tangent to Q is that there exists a scalar $\lambda \neq 0$, and a point $(x : y : z)$ in $\mathbb{P}$ satisfying the following system of linear equations in x, y, z, λ.

$$\begin{cases} \lambda X &= 2(ax + hy + gz) \\ \lambda Y &= 2(hx + by + fz) \\ \lambda Z &= 2(gx + fy + cz) \\ 0 &= Xx + Yy + Zz. \end{cases}$$

It is sufficient to show that this linear system of equations has a non-trivial solution. (Such a solution cannot have $\lambda = 0$, for then automatically $x = y = z = 0$ as the symmetric 3×3 matrix of coefficients in Q is invertible. Thus $\lambda \neq 0$, and hence one of x, y, z is non-zero, else $X = Y = Z = 0$.) By linear algebra, the required condition is that

$$\begin{vmatrix} X & 2a & 2h & 2g \\ Y & 2h & 2b & 2f \\ Z & 2g & 2f & 2c \\ 0 & X & Y & Z. \end{vmatrix} = 0.$$

Expanding the determinant we obtain the following conic $Q^\star$ whose coefficients A, B, ... are the cofactors of a, b, ... in the matrix of coefficients of Q.†

$$Q^\star = Ax^2 + By^2 + Cz^2 + 2Hxy + 2Fyz + 2Gzx.$$

In particular, the matrix of coefficients in $Q^\star$ is (up to a non-zero scalar multiple) the inverse of the matrix of coefficients in Q, so $Q^\star$ is likewise non-singular. The key property of the dual conic is that Q is tangent to a line L if and only if $Q^\star$ passes through the point $L^\star$. Here is an application of these ideas.

Example 16.16 Suppose we are given five distinct lines L_1, L_2, L_3, L_4, L_5 in $\mathbb{P}$ with the property that no three are concurrent, then there is a unique non-singular conic tangent to each of the five lines. In the dual plane $\mathbb{P}^\star$ these five lines correspond to five distinct points $L_1^\star$, $L_2^\star$, $L_3^\star$, $L_4^\star$, $L_5^\star$ with the property that no three are collinear. There is a unique conic

† Recall that for a square matrix $X = (x_{ij})$ the *minor* of the element x_{ij} is the determinant $\det X_{ij}$ of the square matrix X_{ij} obtained from X by deleting the ith row and the jth column; and the *cofactor* of x_{ij} is $(-1)^{(i+j)} \det X_{ij}$.

Q through these five points, necessarily non-singular. And then the dual conic Q^* has the required properties.

Example 16.17 Let Q be a conic in $\mathbb{PC}^2$, and let P be a point which does not lie on Q. In Lemma 12.3, we showed (via the normal form for an irreducible conic) that there are exactly two lines through P tangent to Q. We can easily recover this fact via the dual conic $Q^\star$. Indeed, the lines through P tangent to Q correspond to the two distinct intersections of $Q^\star$ with the line $P^\star$.

Exercises

16.6.1 Find the equation of the dual conic for F in each of the following cases.

(i) $F = xy + yz + zx$
(ii) $F = ax^2 + by^2 + cz^2$ with $abc \neq 0$
(iii) $F = xy + z^2$.

16.6.2 Suppose that in the affine plane $\mathbb{R}^2$ four lines are given, with the property that no two are parallel and no three are concurrent. Show that there exists a unique parabola tangent to each of the four lines.

16.6.3 Let F, G be two conics in $\mathbb{PC}^2$. Show that there are at most four lines tangent both to F and to G.

17

The Group Structure on a Cubic

In this chapter, we will establish the fascinating fact that an irreducible cubic in $P\mathbb{C}^2$ has the natural structure of an Abelian group. This will enable us to describe some of the geometric properties of cubics in group theoretic terms. The starting point is the observation that if P, Q are simple points on an irreducible cubic F, and the line L joining them is not tangent to F at either point, then there is a naturally associated third point $P \star Q$ on F, namely the third point where F meets L. In Section 17.2, we will extend this idea to any two simple points on F, and establish the basic properties of the operation $\star$. The key technical property depends on a subtle result about pencils of cubics, called the Theorem of the Nine Associated Points; the sole object of the next section is to establish this result.

17.1 The Nine Associated Points

By Bézout's Theorem, any two cubic curves in $P\mathbb{C}^2$ having no common component intersect in nine points, counted properly. One of the keys to understanding the geometry of cubic curves is the following result.

Lemma 17.1 *Let F_1, F_2 be cubic curves in* $P\mathbb{C}^2$ *having no common component, and intersecting in nine distinct points $P_1, \ldots, P_9$. Then any cubic F which passes through $P_1, \ldots, P_8$ must pass through P_9.* (Theorem of the Nine Associated Points.)

Proof It is enough to show F lies in the pencil spanned by F_1, F_2. Suppose otherwise. Then F_1, F_2, F span a linear system $\mathscr{L}$ of dimension 2, and (by Lemma 16.6) given any two points Q_1, Q_2 there exists a curve G in $\mathscr{L}$ through Q_1, Q_2. We show, by considering cases, that this leads to a

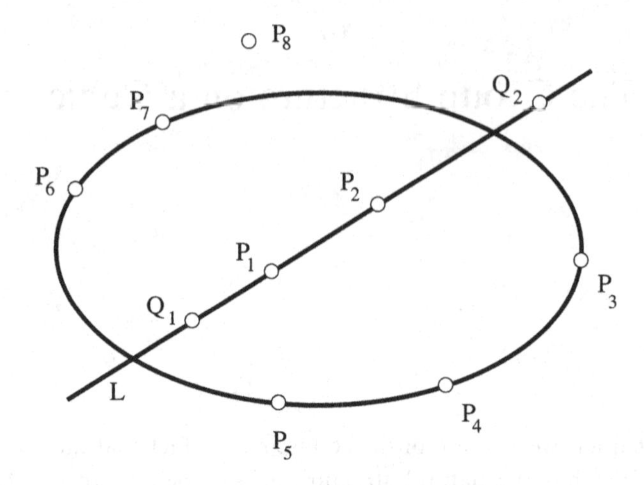

Fig. 17.1. Nine associated points: the general case

contradiction. Note that no four P_i can lie on a line; by Bézout's Theorem that line would be a common component of F_1, F_2 and their intersection would be infinite, contrary to hypothesis. For the same reason, no seven of the P_i can lie on a conic.

Case 1: The 'general' case is when no three of the P_i lie on a line, and no six lie on a conic. (Figure 17.1.) Let L be the line through P_1, P_2, let Q_1, Q_2 to be any two distinct points on L distinct from P_1, P_2, and let C be a conic through P_3, P_4, P_5, P_6, P_7. (Recall that there always exists at least one conic through five points.) We choose G to be a cubic in the linear system spanned by F_1, F_2, F which passes through Q_1, Q_2. By hypothesis F_1, F_2, F all pass through P_1, P_2 so G (which is a linear combination of these three cubics) does so as well. Thus G intersects L in at least four distinct points P_1, P_2, Q_1, Q_2. Since G is a *cubic* that means that L is a component of G, so we can write $G = LC'$ for some conic C'. We claim that C, C' are the same conic. Well, C' passes through P_3, P_4, P_5, P_6, P_7 since G does, and L does not pass through any of these points. (Remember, we are assuming in this case that no three of the P_i lie on a line.) However, C has the same property, so C, C' intersect in at least five distinct points. That can only happen when C, C' have a common component, so either C, C' have a common line component, or they coincide. In the former case, the line component would have to pass through at least three of the P_i (which case we are excluding) so C, C' coincide. Neither L nor C pass through P_8 (remember, we are assuming that no six of the P_i lie on a conic) so $G = LC$ cannot pass through P_8; that is a contradiction, for G must pass through P_8 as F_1, F_2, F do.

Case 2: The points P_1, P_2, P_3 lie on a line L_1 and P_4, P_5, P_6 lie on a line L_2. Choose Q_1 on L_1 distinct from P_1, P_2, P_3, and Q_2 not on L_1, L_2 nor the line joining P_7, P_8. Construct G as above. G passes through P_1, P_2, P_3, Q_1 so has L_1 as a component. The residual conic passes through P_4, P_5, P_6 so has L_2 as a component. Thus P_7, P_8, Q_2 lie on a line, contradicting the choice of Q_2.

Case 3. The points P_1, P_2, P_3 lie on a line L, but no three of P_4, P_5, P_6, P_7, P_8 lie on a line. There exists a conic C through these five points, necessarily irreducible. Choose Q_1 to be a fourth point on L, and Q_2 not lying on L or C. As in Case 1, the resulting G has L as a component, and the residual conic passes through P_4, P_5, P_6, P_7, P_8 so coincides with C. However, that contradicts the choice of Q_2.

Case 4. The points $P_1, \dots, P_6$ lie on a conic C. Choose Q_1 to be a seventh point on C, and Q_2 to be any point not on C, nor on the line joining P_7, P_8. This time G has C as a component, and the residual line passes through P_7, P_8, Q_2 contradicting the choice of Q_2.

□

Example 17.1 Suppose we have six distinct points $V_1, \dots, V_6$ on an irreducible conic C in $\mathbb{PR}^2$. Think of these as the *vertices* of a hexagon (known as Pascal's Mystic Hexagon) inscribed in C, with *sides* the six lines $L_1, \dots, L_6$ joining consecutive points V_1, V_2 and so on up to V_6, V_1. (Figure 17.2.) Pascal's Theorem is a classical result in the geometry of conics, which says that the pairs of opposite sides of the hexagon meet in three (distinct) collinear points. This goes as follows. Let P_{14}, P_{25} and P_{36} be the points of intersection of L_1, L_4, of L_2, L_5 and of L_3, L_6. Note first that none of the P's lie on C, since otherwise we would have three distinct collinear points on C, contradicting the assumption that C is irreducible. Also, no two of the P's can coincide; for instance, if $P_{14} = P_{25} = P$ then V_2, P both lie on the lines L_1, L_2, so these lines would coincide, entailing a coincidence amongst the vertices. Let L be the line joining P_{14}, P_{25}. Then $L_1L_3L_5$ and $L_2L_4L_6$ are cubic curves intersecting in the six vertices of the hexagon and the three P's. However, LC is a cubic curve which passes through eight of these points, namely the six vertices of the hexagon and two of the P's. By the Theorem of the Nine Associated Points, LC must pass through the ninth point P_{36}. But P_{36} cannot lie on the conic C, so must lie on the line L.

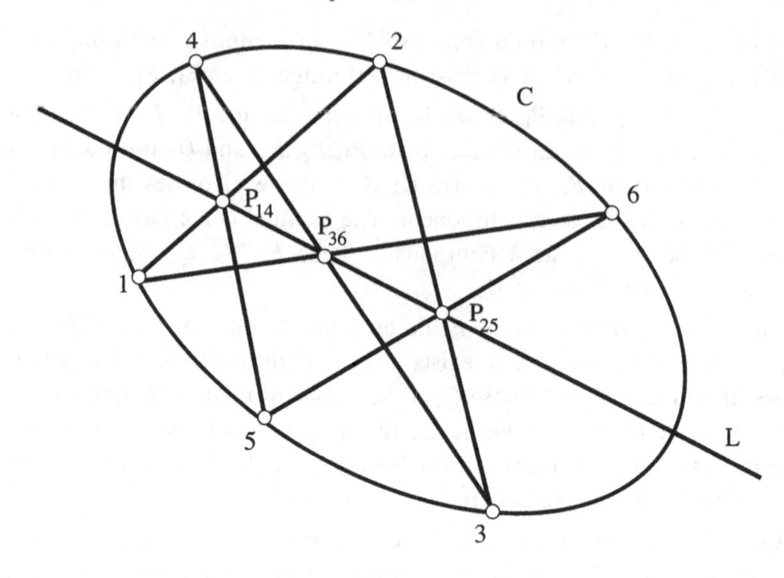

Fig. 17.2. Pascal's Mystic Hexagon

Exercises

17.1.1 Let F, G be curves of degree d in $P\mathbb{C}^2$, intersecting in exactly d^2 distinct points, and let E be an irreducible curve of degree c which passes through exactly cd of these points. Show that there exists a curve in the pencil $\lambda F + \mu G$ which has E as an irreducible component. (Choose a curve in the pencil which passes through an appropriately chosen point on E, and use Bézout's Theorem.) Deduce that the remaining $d(d - c)$ points of intersection lie on a curve of degree $(d - c)$.

17.1.2 Consider the case $d = 3$, $c = 2$ of the previous exercise, when two cubics F, G intersect in nine distinct points, and exactly six of these nine points lie on an *irreducible* conic E. The result is that there is a curve in the pencil having E as an irreducible component. Deduce that the remaining three intersections of F, G are collinear. Use this result to establish Pascal's Theorem, namely that the three pairs of opposite sides of a hexagon inscribed in an irreducible conic E meet in three collinear points.

17.1.3 Show how to modify the argument of Exercise 17.1.2 to deal with the case when the conic E reduces to two distinct lines. Use the result to establish the classical theorem of Pappus, namely

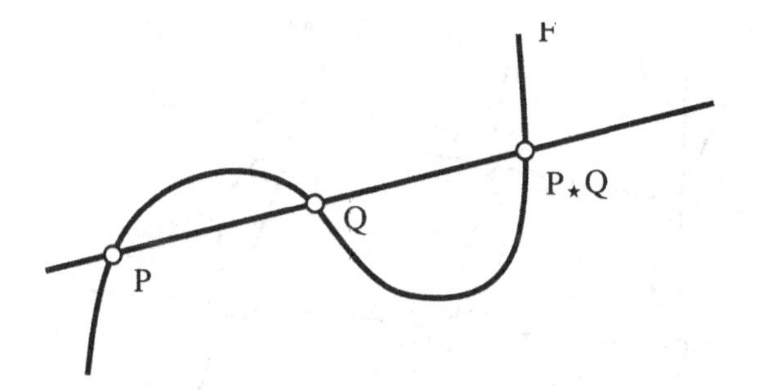

Fig. 17.3. The star operation

that the three pairs of opposite sides of a hexagon inscribed in a line-pair meet in three collinear points. (Draw a diagram.)

17.1.4 Use the Duality Principle to establish the following theorem of Brianchon: if a conic is tangent to each of the six sides of a hexagon then the three lines joining opposite pairs of vertices are concurrent.

17.2 The Star Operation

Let F be an irreducible cubic curve in $P\mathbb{C}^2$, and let P, Q be simple points on F. Suppose first that P, Q are distinct, with the line L joining them not tangent to F at P or Q. There is a naturally associated third point $P \star Q$ on F, namely the third point where F meets L. (Figure 17.3.) We can extend the definition to any two distinct simple points P, Q by agreeing that $P \star Q = P$ when L is tangent to F at P, and that $P \star Q = Q$ when L is tangent to F at Q. It remains to consider the cases when P, Q coincide. Provided P is not a flex, it is then natural to define $P \star Q$ to be the point where the tangent line to F at P meets F again; in the case when P is a flex, we define $P \star Q = P$.

Note that in every case $P \star Q$ is a simple point on F, so $\star$ defines a binary operation on the set S_F of all simple points on F. The key properties of the operation $\star$ are as follows.

Lemma 17.2 *The binary operation $\star$ on S_F has the following basic properties.*

(i) $P \star Q = Q \star P$
(ii) $(P \star Q) \star P = Q$
(iii) $((P \star Q) \star R) \star S = P \star ((Q \star S) \star R)$.

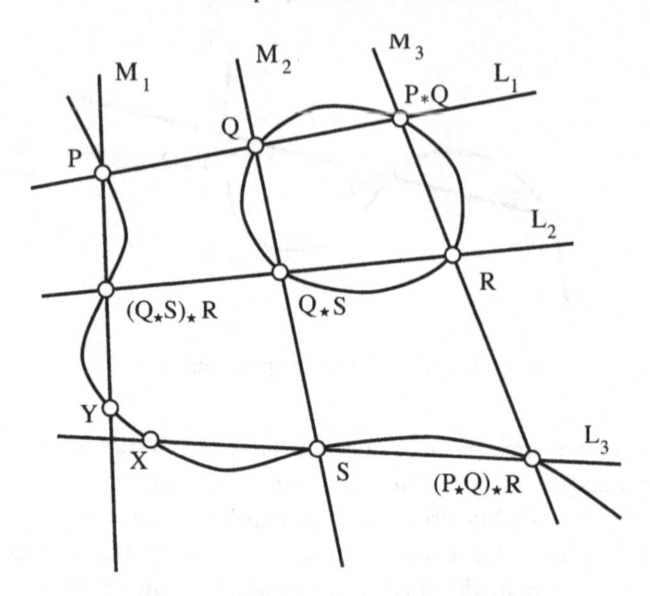

Fig. 17.4. Properties of the star operation

Proof (i), (ii) follow immediately from the definitions. For (iii) let $X =$ LHS, $Y =$ RHS. With the notation of Figure 17.4, the cubic curves $L_1L_2L_3$, $M_1M_2M_3$ meet in nine points and F passes through eight of them. By the Theorem of the Nine Associated Points, F passes through the ninth point, so $X = Y$. □

The perceptive reader may note that (strictly speaking) there is a gap in the above proof. One of the hypotheses in the Theorem of the Nine Associated Points is that the nine points are *distinct*, a condition which is not automatically satisfied in the above proof. The gap can be filled by a 'continuity' argument; however, in this book we do not assume familiarity with topological arguments, so omit it.

17.3 Cubics as Groups

Now let O (the *base point*) be any fixed simple point on an irreducible cubic curve F in $P\mathbb{C}^2$. We define a binary operation $+$ on S_F by $P + Q = (P \star Q) \star O$.

Lemma 17.3 *S_F is a group under the operation $+$, with identity O. The inverse $-S$ of an element S is given by $-S = S \star (O \star O)$.*

Proof It follows from the definitions that the operation $\star$, and hence the operation $+$, is commutative. $+$ is also associative. By the definitions

$$(P+Q)+R = ((P+Q)\star R)\star O = (((P\star Q)\star O)\star R)\star O$$

$$P+(Q+R) = (P\star(Q+R))\star O = (P\star((Q\star R)\star O))\star O$$

and the RHS's are equal by property (iii) in Lemma 17.2. The point O is an identity element for $+$, since for any simple point S we have

$$S+O = (S\star O)\star O = (O\star S)\star O = S.$$

Finally, S has inverse $-S = S\star(O\star O)$: using the displayed relations (i) and (ii) in Lemma 17.2 we have

$$\begin{aligned} S+(-S) &= S+(S\star(O\star O)) \\ &= (S\star(S\star(O\star O)))\star O \\ &= ((S\star(O\star O))\star S)\star O \\ &= (O\star O)\star O = O. \end{aligned}$$

□

The definition of the group structure on S_F depends on the choice of base point O. However the next proposition shows that the isomorphism type of the abstract group does not depend on this choice.

Lemma 17.4 *The groups arising from distinct choices O, O' of base point are isomorphic.*

Proof Let $(S_F,+)$ be the group arising from the base point O, and $(S_F,+')$ the group arising from the base point O'. Set $A = O\star O'$. We claim that the mapping $\Phi : (S_F,+)\to(S_F,+')$ defined by $\Phi(P) = A\star P$ is an isomorphism of groups. First, Φ is injective, for if $\Phi(P)=\Phi(Q)$ then $A\star P = A\star Q$, so $(A\star P)\star A = (A\star Q)\star A$, i.e. $P = Q$. The mapping Φ is also surjective, for given any element R we have $\Phi(R\star A) - A\star(R\star A) = R$. Thus Φ is a bijection. It remains to show that Φ is a group homomorphism. What we have to show is that for all P, Q we have $\Phi(P+Q) = \Phi(P)+'\Phi(Q)$. This we do by straight computation, using the three basic properties of the operation $\star$ listed in Lemma 17.2.

$$\begin{aligned} \Phi(P)+'\Phi(Q) &= \{(A\star P)\star(A\star Q)\}\star O' \\ &= \{(A\star P)\star(Q\star A)\}\star O' \\ &= A\star\{(P\star O')\star(Q\star A)\} \end{aligned}$$

$$\begin{aligned} &= A \star \{(A \star Q) \star (P \star O')\} \\ &= A \star \{((O \star O') \star Q) \star (P \star O')\} \\ &= A \star \{O \star \{(O' \star (P \star O')) \star Q\}\} \\ &= A \star \{O \star (P \star Q)\} \\ &= \Phi(P + Q). \end{aligned}$$

□

In practice, it is advantageous to choose O to be a flex of F, for then some basic geometric facts about the curve F have neat interpretations in terms of the group structure. Note that in that case $O \star O = O$, so the inverse $-P = P \star O$.

Lemma 17.5 *Let P, Q, R be arbitrary points in the group S_F associated to an irreducible cubic F with base point a flex O. Then*

(i) *$P + Q + R = O$ if and only if P, Q, R are collinear.*
(ii) *$P \neq O$ has order 2 if and only if the tangent at P passes through O.*
(iii) *$P \neq O$ has order 3 if and only if P is a flex.*

Proof We will split the proof into three steps, one for each part of the statement.

(i) Suppose that P, Q, R are distinct simple collinear points, so $P \star Q = R$. Then, using the definitions and property (ii) in Lemma 17.2 we obtain $(P + Q) + R = ((P + Q) \star R) \star O = (((P \star Q) \star O) \star R) \star O = ((R \star O) \star R) \star O = O \star O = O$. And conversely, if $P + Q + R = O$ then $P + Q = -R$ so $(P \star Q) \star O = R \star O$ and $(P \star Q) + O = R + O$, i.e. $P \star Q = R$ and P, Q, R are collinear.

(ii) Now let $P \neq O$ be a point on F. Then P has order 2 in the group if and only if $2P = P + P = O$, i.e. if and only if $(P \star P) \star O = O$, i.e. if and only if $P \star P = O$. The result follows, as $P \star P$ is the point on F where the tangent at P meets the curve again.

(iii) Suppose $P \neq O$ and that P is a flex, i.e. $P \star P = P$. Then $P + P = (P \star P) \star O = P \star O \neq O$ (so P cannot be of order 2) and $(P + P) + P = ((P \star O) \star P) \star O = O \star O = O$, so P has order 3. Conversely, suppose $P \neq O$, and P has order 3, i.e. $P + P \neq O$ and $(P + P) + P = O$. Then $P + P = -P$ i.e. $(P \star P) \star O = P \star O$ and hence $P \star P = P$ which means P is a flex.

□

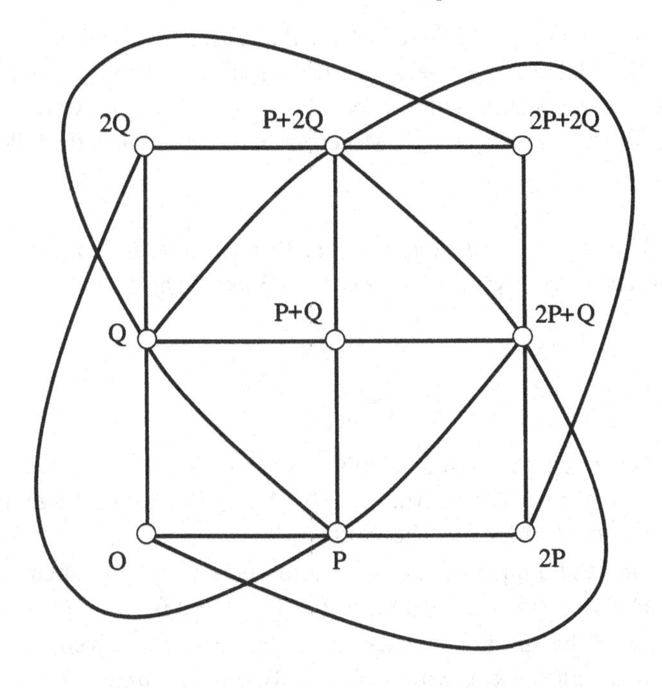

Fig. 17.5. Configuration of flexes on a general cubic

Example 17.2 Let F be an irreducible cubic, with base point a flex O, and let P, Q be distinct flexes on F. (We allow the possibility that one of P, Q coincides with O.) We claim that on the line joining P, Q there is a third flex, namely $2P + 2Q$. This is a straight application of Lemma 17.5. The line meets F in a third point R satisfying $P+Q+R = 0$, so $R = -P - Q$. But P, Q are flexes, so $3P = O$, $3Q = O$. It follows that $-P = 2P$, $-Q = 2Q$ so $R = 2P + 2Q$ and $3R = 6P + 6Q = O + O = O$. It follows that R is a flex.

The argument of Example 17.2 shows that given two distinct flexes P, Q on an irreducible cubic F, there is a third flex $2P + 2Q$ (distinct from the first two) on the line joining them. It also shows that given any three *non-collinear* flexes O, P, Q on an irreducible cubic F, we can construct nine distinct flexes, namely O, P, $2P$, Q, $P + Q$, $2P + Q$, $2Q$, $P + 2Q$, $2P + 2Q$. These form the nine point configuration discussed in Section 13.2 and illustrated in Figure 17.5.

It follows from these considerations that an irreducible cubic F in $P\mathbb{C}^2$ has one, three or nine inflexions. That is entirely consistent with the results of Chapter 15 where we saw that cuspidal, nodal and general cubics in $\mathbb{PC}^2$ have respectively one, three and nine flexes. For instance, the flexes of the Steiner cubic $F = x^3 + y^3 + z^3 + 3\lambda xyz$ were written down explicitly in Example 13.6. Moreover, Lemma 13.6 has the following consequence.

Lemma 17.6 *Any general cubic curve in $\mathbb{PC}^2$ is projectively equivalent to a Steiner cubic $x^3 + y^3 + z^3 + 3\lambda xyz$ for some complex number λ with $\lambda^3 \neq -1$.*

Exercises

17.3.1 Let F be the cuspidal cubic defined by $y^2z = x^3$ with cusp $C = (0 : 0 : 1)$ and flex $O = (0 : 1 : 0)$. Show that there are no points of order 2 in the group S_F.

17.3.2 Find the unique simple point of order 2 on the nodal cubic $F = x^3 + y^3 - xyz$, with node $N = (0 : 0 : 1)$ and flex $O = (1 : -1 : 0)$.

17.3.3 Let F be the non-singular cubic curve in $\mathbb{PC}^2$ given by $y^2z = x(x - z)(x - kz)$, with $k \neq 0, 1$. Taking the base point for the group structure on F to be the flex $O = (0 : 1 : 0)$, determine the four points of order ≤ 2 on F.

17.4 Group Computations

In Chapter 15, we saw that any irreducible cubic in $\mathbb{PC}^2$ is projectively equivalent to a Weierstrass normal form $y^2z = x^3 + ax^2z + bxz^2 + cz^3$. Note that the point $O = (0 : 1 : 0)$ lies on the curve, and is simple; moreover, the line at infinity $z = 0$ meets the curve when $x^3 = 0$, so O is a flex and a natural choice of base point. With that choice, the line through any point $R = (\alpha : \beta : \gamma)$ and O is $\gamma x = \alpha z$, for a point with $\gamma \neq 0$, it can be thought of affinely as the 'vertical' line $x = \alpha/\gamma$ through P. And this line meets the curve again at the inverse $-R = (\alpha : -\beta : \gamma)$, its 'reflexion' in the x-axis, as in Figure 17.6. Thus given simple points P, Q on the curve, the point $P + Q$ is the 'reflexion' of the point $R = P \star Q$ in the x-axis.

Example 17.3 Here is an example of a computation with a Weierstrass normal form F. Let $F = y^2z - x^3 - 4xz^2$, with base point the flex $O = (0 : 1 : 0)$. We claim that $A = (2 : 4 : 1)$ has order 4. What are the

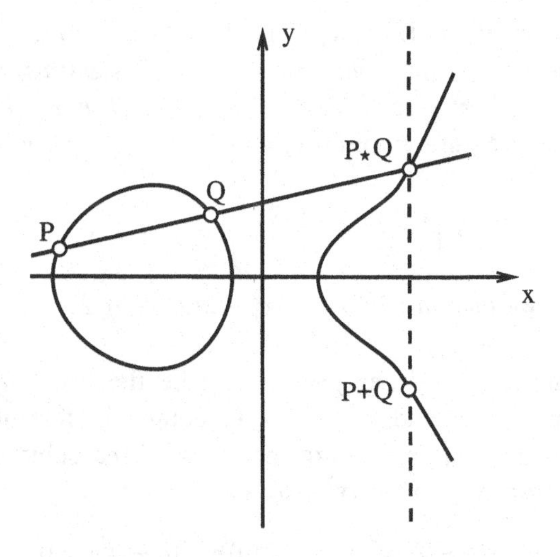

Fig. 17.6. Group operations on a Weierstrass normal form

coordinates of the point $2A$? We have

$$F_x(2,4,1) = -16, \quad F_y(2,4,1) = 8, \quad F_z(2,4,1) = 0.$$

Thus the tangent at A is $-16x + 8y + 0z = 0$ i.e. $y = 2x$. The tangent meets the curve again when $4x^2z = x^3 + 4xz^2$, i.e. $x(x - 2z)^2 = 0$. The factor $x - 2z$ corresponds to the tangency at A. When $x = 0$ we get $y^2z = 0$ so $y = 0$ or $z = 0$; $y = 0$ yields the point $(0 : 0 : 1)$ on $y = 2x$, and $z = 0$ yields the point $O = (0 : 1 : 0)$ *not* on $y = 2x$. Thus $A \star A = (0 : 0 : 1)$. Reflexion in the x-axis yields $2A = (0 : 0 : 1)$. Next, we compute $3A = A + 2A$. The line through A, $2A$ is the tangent $y = 2x$ at A, so $A \star 2A = A$, and reflexion in the x-axis yields $3A = (2 : -4 : 1)$. Finally, we compute $4A = A + 3A$. The line through A, $3A$ is the 'vertical' line $x = 2z$, which meets the curve again at $O = (0 : 1 : 0)$, so $A \star 3A = O$, and reflexion in the x-axis yields the same point $O = (0 : 1 : 0)$. Thus $4A = 0$, establishing that A has order 4, so generates the cyclic group of order 4.

Doing such computations on a case-by-case basis is clearly going to be inefficient. The point of the next proposition is that it mechanizes this kind of computation for Weierstrass normal forms, by producing explicit formulas for the group operation. (In particular, such computations can be programmed on a computer.)

Lemma 17.7 *Let $P_1 = (x_1 : y_1 : 1)$, $P_2 = (x_2 : y_2 : 1)$ and $P_k = (x_3 : y_3 : 1)$ be three simple points on a cubic in Weierstrass normal form $y^2z = x^3 + ax^2z + bxz^2 + cz^3$ with $P_3 = P_1 + P_2$. Then, provided $x_1 \neq x_2$, the coordinates of P_3 are given by $x_3 = \lambda^2 - a - x_1 - x_2$, $y_3 = -\lambda x_3 - \mu$ where*

$$\lambda = \frac{y_1 - y_2}{x_1 - x_2}, \qquad \mu = y_1 - \lambda x_1.$$

When $x_1 = x_2$, we take instead $\lambda = (3x_1^2 + 2ax_1 + b)/2y_1$.

Proof Assume $x_1 \neq x_2$. Let $y = \lambda x + \mu$ be the line joining P_1, P_2. Substituting $y = y_1, y_2$ and $x = x_1, x_2$ we obtain the formulas for λ, μ. Further, the scalars x_1, x_2, x_3 are the roots of the cubic obtained by eliminating y from $y^2 = x^3 + ax^2 + bx + c$, $y = \lambda x + \mu$, i.e. the cubic

$$x^3 + (a - \lambda^2)x^2 + (b - 2\lambda\mu)x + (c - \mu^2) = 0.$$

By the Newton Relations $x_1 + x_2 + x_3 = \lambda^2 - a$, giving the formula for x_3. The formula for y_3 is immediate, since $-y_3 = \lambda x_3 + \mu$. (The line joining O to a point on the curve is the 'vertical' line through that point, and meets the curve again at the reflexion of that point in the x-axis.) It remains to consider the case $x_1 = x_2$. Either $P_2 = -P_1$ or $P_1 = P_2$. In the former case $P_1 + P_2 = O$, contradicting the hypothesis that P_3 is a finite point. In the latter case, the tangent at $P = P_1 = P_2$ has equation $y = \lambda x + \mu$, with $\lambda = F_x/F_y = (3x_1^2 + 2ax_1 + b)/2y_1$, and the formulas for x_3, y_3 are as before. □

Example 17.4 Let F be the non-singular cubic curve in $\mathbb{PC}^2$ with equation $y^2z = x^3 - 43xz^2 + 166z^3$. We take $O = (0 : 1 : 0)$ as the base point for the group structure. We will verify that $P = (3 : 8 : 1)$ is an element of order 7 in F. For this example, the formulas of the above proposition become $x_3 = \lambda^2 - x_1 - x_2$ and $y_3 = -\lambda x_3 - \mu$ where for $P_1 \neq P_2$ we have

$$\lambda = \frac{y_1 - y_2}{x_1 - x_2}, \qquad \mu = y_1 - \lambda x_1 = y_2 - \lambda x_2.$$

And when $P_1 = P_2$ the formula is modified by writing $\lambda = (3x_1^2 - 43)/2y_1$. The reader is left to verify that

$$\begin{array}{lll} 2P = (-5 : -16 : 1) & 3P = (11 : -32 : 1) & 4P = (11 : 32 : 1) \\ 5P = (-5 : 16 : 1) & 6P = (3 : -8 : 1) & 7P = (0 : 1 : 0) = O. \end{array}$$

Exercises

17.4.1 Let F be the non-singular cubic curve in $\mathbb{PC}^2$ defined by $y^2z = x^3 + z^3$. We take the flex $O = (0:1:0)$ as the base point for the group structure on S_F. Show that $P = (2:3:1)$ is an element of order 6 in F.

17.4.2 Let F be the cubic curve in $\mathbb{PC}^2$ defined by $y^2z + 7xyz = x^3 + 16xz^2$. Show that F is non-singular. Show also that $O = (0:1:0)$ is a flex on F, which we take as the base point for the group structure on S_F. Show that $P = (2:2:1)$ is an element of order 8 in S_F.

17.4.3 Let F be the cubic curve in $\mathbb{PC}^2$ with equation $y^2z + xyz = x^3 - 45xz^2 + 81z^3$. Show that F is non-singular. Show also that $O = (0:1:0)$ is a flex on F, which we take as the base point for the group structure. Show that $P = (0:9:1)$ is an element of order 10 in S_F.

17.5 Determination of the Groups

The object of this section is to determine the group of an irreducible cubic in $\mathbb{PC}^2$, up to isomorphism. We will achieve this for cubics of cuspidal and nodal types, and simply state the result for cubics of general type, since the proof requires a substantial excursion into complex function theory. The first thing to be clear about is that the isomorphism type of the group depends only on the projective equivalence class of the curve.

Lemma 17.8 *Let F, G be irreducible cubics in $\mathbb{PC}^2$. If F, G are projectively equivalent then the associated groups are isomorphic.*

Proof Let Φ be a projective equivalence mapping the zero set of F bijectively to the zero set of G. In particular, Φ maps the set F' of simple points on F bijectively to the set G' of simple points on G. It is enough to show that Φ is a group homomorphism, hence a group isomorphism.

Our first claim is that $\Phi(P \star Q) = \Phi(P) \star \Phi(Q)$ for all simple points P, Q on F. Consider first the case when P, Q are distinct and the line L joining them is tangent at neither point, thus $P \star Q$ is the third point where L meets F. Then Φ maps L to the line $\Phi(L)$ through $\Phi(P)$, $\Phi(Q)$. Since $\Phi(L)$ meets F again at the point $\Phi(P \star Q)$ the claim is immediate in this case. Suppose, however, that L is tangent to F at P (respectively Q) so that $P \star Q = P$ (respectively $P \star Q = Q$). Then $\Phi(L)$ is tangent to G

at $\Phi(P)$ (respectively $\Phi(Q)$), and it is immediate that $\Phi(P \star Q) = \Phi(P) = \Phi(P) \star \Phi(Q)$ (respectively $\Phi(P \star Q) = \Phi(Q) = \Phi(P) \star \Phi(Q)$). Now assume that $P = Q$ (but that point is not a flex) and let T be the tangent line to F at that point, so $P \star Q$ is the second point where T meets F. Then $\Phi(P) = \Phi(Q)$, $\Phi(T)$ is the tangent line to G at that point, and $\Phi(P \star Q)$ is the point where the tangent meets G again, thus $\Phi(P \star Q) = \Phi(P) \star \Phi(Q)$. Finally, suppose that $P = Q$ is a flex. Then (again, since intersection numbers with lines are invariant under projective mappings) $\Phi(P) = \Phi(Q)$ is a flex, and $\Phi(P \star Q) = \Phi(P) = \Phi(P) \star \Phi(Q)$.

It is now an easy matter to show that Φ is a group homomorphism, i.e. that $\Phi(P + Q) = \Phi(P) + \Phi(Q)$ for all simple points P, Q on F. Choose a base point O_F on F, and take the base point O_G on G to be $O_G = \Phi(O_F)$. Then $\Phi(P + Q) = \Phi((P \star Q) \star O_F) = \Phi(P \star Q) \star \Phi(O_F) = (\Phi(P) \star \Phi(Q)) \star O_G = \Phi(P) + \Phi(Q)$. □

We are now in a position to determine the groups associated to cubics of cuspidal and nodal type.

Lemma 17.9 *Let F be a cuspidal cubic in $P\mathbb{C}^2$. Then the associated group S_F is isomorphic to the additive group $\mathbb{C}$ of complex numbers.*

Proof By Lemma 17.8 it suffices to work with the normal form $F = y^2z - x^3$ for the cuspidal cubic in $P\mathbb{C}^2$, with cusp $C = (0 : 0 : 1)$ and unique flex $O = (0 : 1 : 0)$ which we take as the base point. Note that in the affine view given by $y = 1$, the point C is the only point at infinity on F. Thus we can identify S_F with the zero set of the affine curve $f = z - x^3$. The points of this affine cubic are parametrized as $x = t$, $z = t^3$, so the points of S_F are those of the form $P_t = (t : 1 : t^3)$ for some complex number t, and we have a natural bijective mapping $\phi : \mathbb{C} \to S_F$ given by $\phi(t) = P_t$. Note that $P_0 = O$. In terms of the affine view, inversion corresponds to central reflexion of the affine plane in the origin. We will show that $\mathbb{C}$ (with its usual additive group structure) is isomorphic to S_F (with the additive group structure described above) by proving that ϕ is a group homomorphism, hence a group isomorphism. We claim that $\phi(s) \star \phi(t) = \phi(-s - t)$ for all $s, t \in \mathbb{C}$. Given that, we obtain the desired result, namely that

$$\begin{aligned}\phi(s) + \phi(t) &= \{\phi(s) \star \phi(t)\} \star O = \{\phi(s) \star \phi(t)\} \star \phi(0) \\ &= \phi(-s - t) \star \phi(0) = \phi(s + t).\end{aligned}$$

To establish the claim note first that the points $\phi(s)$, $\phi(t)$, $\phi(-s - t)$

are always collinear, by the determinantal criterion of Example 9.2. That establishes the claim when the parameters s, t, $-s-t$ are distinct. To deal with the cases when two of these coincide, observe that the tangent to F at any point P_t is $-3t^2x+2t^3y+z$, which passes through P_{-2t}; the claim follows. □

Example 17.5 The algebraic result of Lemma 17.9 is consistent with our geometric results about cuspidal cubics. For instance there are no elements of finite order (apart from the identity) in the additive group $\mathbb{C}$, so the same must be true of the group S_F. In particular, there are no elements of order 2, so we deduce that there are no points $P \neq O$ where the tangent passes through O, nor are there any points of order 3, so there are no flexes $P \neq O$, confirming the result of Exercise 13.2.

Lemma 17.10 *Let F be a nodal cubic in $P\mathbb{C}^2$. Then the associated group S_F is isomorphic to the group $\mathbb{C}^*$ of non-zero complex numbers under the operation of multiplication.*

Proof We follow the previous proof very closely. By Lemma 17.8, it suffices to work with the normal form $F = x^3 + y^3 - xyz$ associated to a nodal cubic in $P\mathbb{C}^2$, with unique singular point $P = (0:0:1)$, and inflexions $(-1:1:0)$, $(-\omega:\omega^2:0)$, $(-\omega^2:\omega:0)$ where ω is a primitive complex cube root of unity. (Example 13.3.) We take the base point to be $O = (-1:1:0)$. The points of S_F are parametrized as $x = -t$, $y = t^2$, $z = 1 - t^3$ with $t \neq 0$, so there is a mapping $\Phi : \mathbb{C}^* \to S_F$ defined by $\Phi(t) = P_t$ where $P_t = (-t, t^2, 1 - t^3)$. Clearly, Φ is a bijection, with $\Phi(1) = O$. We will show that $\mathbb{C}^*$ (with its usual multiplicative group structure) is isomorphic to S_F (with the additive group structure described above) by proving that Φ is a homomorphism, hence a group isomorphism. We claim that $\phi(s) \star \phi(t) = \phi(1/st)$ for all $s, t \in \mathbb{C}^*$. Given that we obtain the desired result, namely that

$$\begin{aligned}\phi(s) + \phi(t) &= \{\phi(s) \star \phi(t)\} \star O = \{\phi(s) \star \phi(t)\} \star \phi(1) \\ &= \phi(1/st) \star \phi(1) = \phi(st)\end{aligned}$$

To establish the claim, note first that the points $\phi(s)$, $\phi(t)$, $\phi(1/st)$ are collinear, by the determinantal criterion. That establishes the claim when the parameters s, t, $1/st$ are distinct. To deal with the cases when two of these coincide, observe that the tangent to F at P_t is $(2t^2 + t^5)x + (2t^4 + t)y + t^3z = 0$, which passes through P_{1/t^2}. □

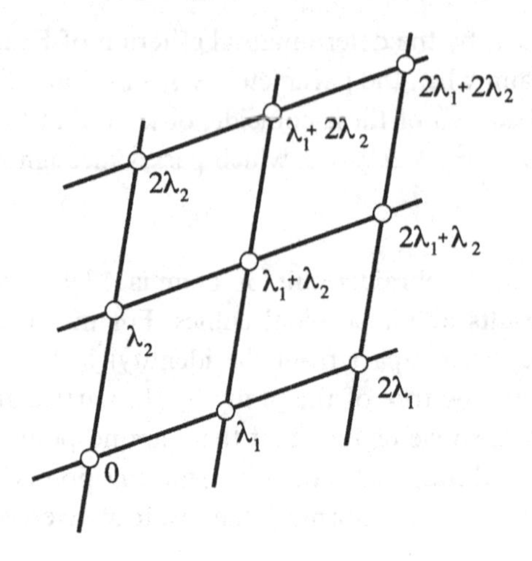

Fig. 17.7. A lattice in $\mathbb{C}^2$

Example 17.6 The algebraic result of Lemma 17.10 is also consistent with our geometric results about nodal cubics. For instance, there are exactly n elements of finite order n in the multiplicative group $\mathbb{C}^*$, namely the complex n^{th} roots of unity, so the same must be true of the group S_F. In particular, there are exactly three elements of order 3, so we deduce that there are exactly three flexes, confirming the result of Example 13.3.

That leaves us with the most interesting case of all, namely the groups associated to general cubics.

Lemma 17.11 *Let F be a general cubic F in* $P\mathbb{C}^2$. *Then the associated group is isomorphic to a quotient group $\mathbb{C}^2/\Lambda$ where Λ is a subgroup of the additive group $\mathbb{C}^2$ comprising all complex numbers of the form $n_1\lambda_1 + n_2\lambda_2$, where λ_1, λ_2 are fixed non-zero complex numbers with non-real quotient.*

The proof of Lemma 17.11 lies beyond the scope of this text, requiring the machinery of 'doubly periodic' functions in complex function theory. Such a subgroup $\Lambda \subset \mathbb{C}^2$ is referred to as a *lattice* in $\mathbb{C}^2$. Thus the group can be thought of as a parallelogram with opposite sides identified, so is topologically a torus. (Figure 17.7.)

That is part and parcel of a continuing story in the subject, which commenced in Example 2.13, namely that curves in $P\mathbb{C}^2$ can be thought of as real surfaces. Here is some purely heuristic reasoning, which can be formalized. Consider the Steiner cubic $F = x^3 + y^3 + z^3 + 3\lambda xyz$ in $P\mathbb{C}^2$.

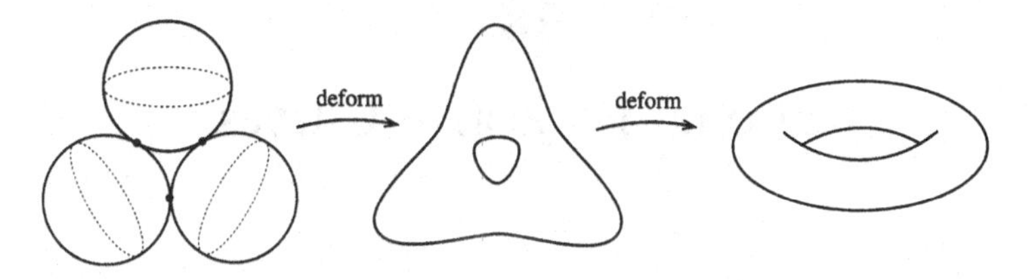

Fig. 17.8. Deforming a triangle to obtain a torus

In Example 10.8, we showed that F is singular if and only if $\lambda^3 = -1$, and that in that case F reduces to a triangle. Consider then the case $\lambda = -1$ when F reduces to three lines. As in Example 9.7, each of these three (complex projective) lines represents a sphere (albeit in real 4-space) with each pair of spheres intersecting in just one point. (Figure 17.8.) Changing the value of λ slightly deforms this picture: what happens is that each common point of two spheres develops into a smooth 'neck' between them, producing a single smooth surface. It does not require too much imagination to see that this surface can be further deformed into a 'torus', a surface with a single hole in it.

18

Rational Projective Curves

In this final chapter, we return to the question of the 'rationality' of curves, which we touched upon briefly (in the affine case) in Chapter 8. It is more natural to study the concept in the projective case, since the behaviour of the curve 'at infinity' plays a role. Ideally, one seeks a necessary and sufficient condition for a curve to be rational. We would need to develop rather more algebra to expose the elegant answer to this question provided by curve theory. Our compromise is to show that a useful class of curves (those of 'deficiency' zero) are rational, and that a substantial class of curves (the non-singular ones of degree ≥ 3) fail to be rational. These results require most of the machinery we have developed in this text, and provide good illustrations of the underlying techniques.

18.1 The Projective Concept

The idea of a 'rational' curve can be extended from the affine plane to the projective plane in a very natural way. An irreducible projective curve F of degree d in $\mathrm{P}\mathbb{K}^2$ is *rational* when there exist forms $X(s,t)$, $Y(s,t)$, $Z(s,t)$ of the same degree d for which the following conditions hold.

(i) For all but finitely many values of the ratio $(s : t)$, at least one of the scalars $X(s,t)$, $Y(s,t)$, $Z(s,t)$ is non-zero, and we have $F(X(s,t), Y(s,t), Z(s,t)) = 0$.

(ii) For all but finitely many *exceptional* points $(x : y : z)$ with $F(x,y,z) = 0$, there is a unique ratio $(s : t)$ for which $\lambda x = X(s,t)$, $\lambda y = Y(s,t)$, $\lambda z = Z(s,t)$ for some $\lambda \neq 0$.

The first step is to establish that this definition is consistent with the one given in the affine case in Chapter 8.

Lemma 18.1 *Let f be a curve of degree d in $\mathbb{K}^2$, and let F be the corresponding curve of degree d in $P\mathbb{K}^2$. Then f is rational if and only if F is rational.*

Proof Here is an outline, with the details suppressed. Suppose F is rational, so there exist forms $X(s,t)$, $Y(s,t)$, $Z(s,t)$ of degree d with the properties in (i), (ii) above. We leave the reader to check that a rational parametrization for f is provided by

$$x(t) = \frac{X(1,t)}{Z(1,t)}, \qquad y(t) = \frac{Y(1,t)}{Z(1,t)}.$$

Conversely, suppose f is rational, so there exist rational functions $x(t)$, $y(t)$ with the properties (i), (ii) in Section 8.1. We can suppose that $x(t) = a(t)/c(t)$, $y(t) = b(t)/c(t)$ for some polynomials $a(t)$, $b(t)$, $c(t)$ of respective degrees α, β, γ giving rise to the set of points $(x : y : 1)$ in $P\mathbb{K}^2$ with $x = a(t)$, $y = b(t)$, $z = c(t)$. Let $A(s,t)$, $B(s,t)$, $C(s,t)$ be the forms defined by $A(s,t) = s^\alpha a(t/s)$, $B(s,t) = s^\beta b(t/s)$, $C(s,t) = s^\gamma c(t/s)$. Write δ for the maximum of α, β, γ. Then a rational parametrization for F is

$$X(s,t) = s^{\delta-\alpha}A(s,t), \quad Y(s,t) = s^{\delta-\beta}B(s,t), \quad Z(s,t) = s^{\delta-\gamma}C(s,t).$$

□

In practice, replace t by t/s in $x = a(t)$, $y = b(t)$, $z = c(t)$, and multiply through by an appropriate power of s to clear denominators.

Example 18.1 In Example 8.2 we saw, by considering the pencil of lines through the point $p = (-1,0)$, that a rational parametrization for the unit circle $f = x^2 + y^2 - 1$ in $\mathbb{R}^2$ is

$$x = \frac{1-t^2}{1+t^2}, \qquad y = \frac{2t}{1+t^2}.$$

Here $a(t) = 1 - t^2$, $b(t) = 2t$, $c(t) = 1 + t^2$. Replacing t by t/s, and multiplying through by s^2 we obtain the rational parametrization $X = s^2 - t^2$, $Y = 2st$, $Z = s^2 + t^2$ for the projective curve $F = x^2 + y^2 - z^2$. An advantage of working projectively is that now every point on the curve arises from a ratio $(s : t)$. The point $p = (-1,0)$ in $\mathbb{R}^2$ corresponds to $P = (-1 : 0 : 1)$ in $P\mathbb{R}^2$, arising from the ratio $(0 : 1)$.

Exercises

18.1.1 Supply the missing detail in the proof of Lemma 18.1.

18.1.2 In Chapter 8, a number of explicit rational parametrizations of affine curves f were obtained, both in the text and in the exercises. Use these examples to practise the technique of obtaining rational parametrizations for the corresponding projective curve F.

18.1.3 Let F be a rational curve in $\mathbb{PK}^2$, projectively equivalent to the curve G. Show that G is likewise rational.

18.2 Quartics with Three Double Points

In Chapter 8, we met a simple technique for establishing the rationality of a number of affine curves. First, we saw that irreducible conics are rational, by considering the intersections of the conic with the pencil of lines through some fixed point on the conic. Likewise, we saw that irreducible cubics with a single double point are rational, by considering the intersections with the pencil of lines through the singular point. Indeed, the same technique shows that more generally, an irreducible curve of degree $d \geq 3$ in $\mathbb{K}^2$ with a singular point of multiplicity $(d-1)$ is rational. (Exercise 8.1.14.) The object of this section is to push these ideas one stage further, by looking at a special class of examples, namely irreducible quartics with three double points, where rationality can be established via a new idea. This will lead to the concept of the 'deficiency' of a curve, an integer invariant introduced in Section 18.3. Conics, cubics with just one double point, and quartics with three double points all turn out to be examples of curves of 'deficiency' zero, and we will show in Section 18.4 that such curves are rational. The starting point is to recall that in Chapter 11 we established the following proposition about irreducible quartics in $\mathbb{PC}^2$ with three double points.

Lemma 18.2 *Any irreducible quartic curve F in $\mathbb{PC}^2$ having three distinct double points A, B, C is projectively equivalent to a 'prenormal form'*

$$F = ay^2z^2 + bz^2x^2 + cx^2y^2 + 2xyz(fx + gy + hz)$$

with $abc \neq 0$, and having double points at the vertices of reference.

Since the property of rationality is invariant under projective equivalence, it suffices to consider quartics in this prenormal form. The key idea is that of the *quadratic transformation* of $\mathbb{PC}^2$ given by the rule $(x : y : z) \mapsto (1/x : 1/y : 1/z)$. Of course, this only makes sense when $xyz \neq 0$, i.e. provided we avoid points on the sides $x = 0$, $y = 0$, $z = 0$ of

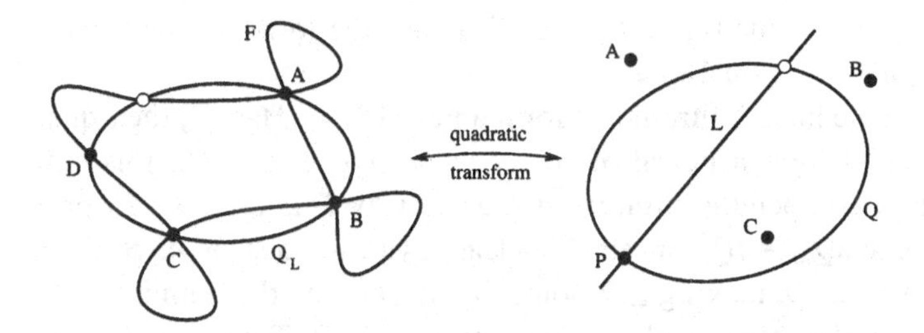

Fig. 18.1. The quadratic transformation

the triangle of reference, but with that proviso it defines a bijective mapping with inverse given by the same formula. Observe that F is obtained from the conic

$$Q = ax^2 + by^2 + cz^2 + 2fyz + 2gzx + 2hxy$$

by applying the quadratic transformation, and then multiplying through by $x^2y^2z^2$. (Note, incidentally, that Q must be irreducible, for if Q were to reduce to two lines, the construction shows that F would reduce to two conics.) Similarly, the conic Q is obtained from the quartic F by applying the quadratic transformation, and then multiplying through by $x^2y^2z^2$. For this reason, we refer to F, Q as *quadratic transforms* of each other. On the basis of this observation alone we can establish

Lemma 18.3 *Any irreducible quartic F in $\mathbb{PC}^2$ having three distinct double points A, B, C is rational.*

Proof By the above comments, it suffices to show that the prenormal form F is rational. We know that the irreducible conic Q obtained via the quadratic transformation has a rational parametrization $X(s,t)$, $Y(s,t)$, $Z(s,t)$. The reader is left to check that a rational parametrization for F is given by $Y(s,t)Z(s,t)$, $Z(s,t)X(s,t)$, $X(s,t)Y(s,t)$. □

The quadratic transform provides the following insight into the geometry. Recall that we can establish rationality of Q by considering the pencil of lines L through a fixed point P on Q, and meeting Q in a second 'variable' point. Now apply the quadratic transformation to this situation. By analogy, we can construct the quadratic transform of a line $L = \alpha x + \beta y + \gamma z$ by applying the quadratic transformation and multiplying through by xyz to obtain a conic $Q_L = \alpha yz + \beta zx + \gamma xy$, passing through A, B, C and the fixed point D on F corresponding to P.

Clearly, *any* conic through A, B, C, D has the form Q_L for some line L through P. (Figure 18.1.)

Since the lines L through P form a pencil $L = sM + tN$, their quadratic transforms form a pencil of conics $Q_L = sQ_M + tQ_N$. On this basis, we obtain a compelling intuitive idea as to why F is rational. In principle, the conic $sQ_M + tQ_N$ meets F at least twice at each of A, B, C, and at least once at D, making ≥ 7 points in all. On the other hand, by Bézout's Theorem the total number of intersections is 8. Thus, *apart from A, B, C, D, the conic $sQ_M + tQ_N$ meets F at most once more* at some variable point $(X : Y : Z)$, whose coordinates depend solely on the ratio $(s : t)$. In any given example one can check that X, Y, Z depend *rationally* on $(s : t)$, yielding an explicit rational parametrization of F. (This kind of reasoning will be formalized in Section 18.4.) An unnatural aspect is that there are infinitely many choices for the point D, corresponding to the infinite number of choices for P. This can be reduced to a finite number of more natural choices by imagining D coinciding with one of the singular points (A say) and interpreting the condition (that the conics pass through D) as the condition that the conics are tangent at A *to a fixed tangent at that point*. With this interpretation, we still have a pencil of conics (Exercise 18.2.2) and one can proceed as before.

Example 18.2 The quartic $F = x^2y^2 - (x^2+y^2)z^2$ in $\mathbb{PC}^2$ has nodes at $A = (1:0:0)$, $B = (0:1:0)$, $C = (0:0:1)$ with tangents $y = \pm z$, $z = \pm x$, $y = \pm ix$. A general conic has the form $ax^2 + by^2 + cz^2 + dxy + eyz + fzx$, and passes through A, B, C if and only if $a = 0$, $b = 0$, $c = 0$, so the required linear system comprises conics of the form $G = dxy + eyz + fzx$. Let $\mathscr{L}$ be the linear system of conics through A, B, C tangent (for instance) to $y = z$ at A. The tangent to G at A is easily checked to be the line $dy + fz = 0$, and this coincides with $y = z$ if and only if $d + f = 0$. Thus $\mathscr{L}$ is the pencil of conics $dx(y-z) + eyz$. At this stage, it is best to work affinely by setting $z = 1$, and writing $t = -e/d$. The pencil is then $x(y-1) = ty$, which we can write better as $y = x/(x-t)$, and our curve is $f = x^2y^2 - (x^2+y^2)$. Substituting for y in this we get $x^2(2xt - t^2 - 1) = 0$. $x = 0$ yields the node C. The variable point of intersection is given by

$$x = \frac{1+t^2}{2t}, \quad y = \frac{x}{x-t} = \frac{1+t^2}{1-t^2}.$$

That is a rational parametrization for f, and yields the following rational parametrization for F:

$$X = s^4 - t^4, \quad Y = 2st(s^2+t^2), \quad Z = 2st(s^2-t^2).$$

Note that A corresponds to the ratios $(1 : 0)$ and $(0 : 1)$, B corresponds to the ratios $(1 : 1)$ and $(1 : -1)$, and C corresponds to the ratios $(i : 1)$ and $(i : -1)$; every other point on F corresponds to a *unique* ratio $(s : t)$.

Example 18.3 We are now in a position to throw some light on the technique used in Example 8.7 to show that the irreducible quartic $f = xy + (x^2 + y^2)^2$ in $\mathbb{R}^2$ is rational. The technique was to consider the intersections of f with the pencil of all circles, centred on the y-axis, and passing through the origin. To see how this fits into the above discussion, consider the associated complex projective curve $F = xyz^2 + (x^2 + y^2)^2$. The reader will readily verify that F has three double points, at $O = (0 : 0 : 1)$, $I = (1 : i : 0)$ and $J = (1 : -i : 0)$. According to the above, we should consider the linear system of all conics through O, I, J. At this point, recall (Exercise 9.5.4) that real irreducible conics through O, I, J are *circles* through O. Also, the tangents to F at O are $x = 0$, $y = 0$ and the condition for the circle to be tangent to the choice $y = 0$ is just that its centre lies on the y-axis. Thus from the complex projective viewpoint it was natural to consider the intersections of f with circles through the origin, centred on the y-axis. In Example 8.7 we obtained the rational parametrization $x = -t^3/(1+t^4)$, $y = t/(1+t^4)$ for f, producing $X = -s^3t$, $Y = st^3$, $Z = s^4 + t^4$ for F.

Exercises

18.2.1 Show that the irreducible quartic $F = (x^2 - z^2)^2 - y^2(2yz + 3z^2)$ in $P\mathbb{C}^2$ has double points at $A = (1 : 0 : 1)$, $B = (-1 : 0 : 1)$, $C = (0 : -1 : 1)$ and a simple point at $D = (0 : 1 : 0)$. By considering the pencil of conics through A, B, C, D, show that F is rational.

18.2.2 Let F be an irreducible quartic in $P\mathbb{C}^2$ with three double points A, B, C, and let L be a tangent at A. Show that the linear system $\mathcal{L}$ of all conics through A, B, C and tangent at A to L is a pencil.

18.2.3 Show that the quartic $F = (x^2 + y^2 - zx)^2 - (x^2 + y^2)z^2$ in $P\mathbb{C}^2$ has ordinary cusps at $O = (0 : 0 : 1)$, $I = (1 : i : 0)$, $J = (1 : -i : 0)$. Find the cuspidal tangent to F at 0. By considering the pencil of all conics through O, I, J and tangent at O to the cuspidal tangent, obtain a rational parametrization for F.

18.2.4 Show that the irreducible quartic $F = (x + z)(x + 2z)y^2 - x^2z^2$

in $\mathbb{PC}^2$ has double points at $A = (1 : 0 : 0)$, $B = (0 : 1 : 0)$, $C = (0 : 0 : 1)$. Obtain a rational parametrization for F.

18.3 The Deficiency of a Curve

Our immediate object is to introduce a projective invariant of a curve, known as the 'deficiency', in order to establish a sufficient condition for rationality in Section 18.4. Before defining the concept, we will establish the following result, which represents an improvement on Lemma 14.10.

Lemma 18.4 *Let F be an irreducible curve of degree d in $\mathbb{PC}^2$, having singular points $X_1, \dots, X_s$ of respective multiplicities $m_1, \dots, m_s$, then*

$$\sum_{k=1}^{s} \frac{1}{2} m_k(m_k - 1) \leq \frac{1}{2}(d-1)(d-2).$$

Proof The result is trivially true for $d = 1, 2$ so we can suppose $d \geq 3$. We split the proof into three steps. Set $M = \sum \frac{1}{2} m_k(m_k - 1)$. Note first that by Lemma 14.10 we have

$$M \leq \frac{1}{2}d(d-1) < \frac{1}{2}(d-1)(d+2) \qquad (*)$$

Step 1: Let $\mathscr{L}$ be the linear system of curves of degree $(d-1)$ which pass through $X_1, \dots, X_s$ with respective multiplicities $\geq m_1 - 1, \dots, \geq m_s - 1$, and let $l = \dim \mathscr{L}$. Then $\operatorname{cod} \mathscr{L} \leq M$ by Lemma 16.5. Since the dimension of the space of curves of degree $(d-1)$ is $\frac{1}{2}(d-1)(d+2)$ we have

$$l = \frac{1}{2}(d-1)(d+2) - \operatorname{cod} \mathscr{L} \geq \frac{1}{2}(d-1)(d+2) - M > 0. \qquad (**)$$

Step 2: The conclusion of Step 1 is that l is positive. Choose l points $Y_1, \dots, Y_l$ on F distinct from the singular points $X_1, \dots, X_s$. (That is possible, since the zero set of a complex projective curve is infinite.) Note that the Y's are necessarily simple on F. Consider now the linear system $\mathscr{L}'$ of all curves of degree $(d-1)$ which pass through $X_1, \dots, X_s$ with respective multiplicities $\geq m_1 - 1, \dots, \geq m_s - 1$, and which also pass through $Y_1, \dots, Y_l$. Also let $\mathscr{L}_k$ be the linear system of curves of degree $(d-1)$ which pass through Y_k. Then $\mathscr{L}' = \mathscr{L} \cap \mathscr{L}_1 \cap \dots \cap \mathscr{L}_l$, and by subadditivity of the codimension we have

$$\operatorname{cod} \mathscr{L}' \leq \operatorname{cod} \mathscr{L} + \operatorname{cod} \mathscr{L}_1 + \cdots + \operatorname{cod} \mathscr{L}_l = \operatorname{cod} \mathscr{L} + l.$$

Thus (using Step 1) we see that $\dim \mathscr{L}' \geq \dim \mathscr{L} - l = 0$. It follows that $\mathscr{L}'$ contains at least one curve F' of degree $(d-1)$.

Step 3: F, F' have no common component, since F is assumed irreducible, and F' has smaller degree than F. Using Bézout's Theorem, the Multiplicity Inequality and (**) we obtain the following inequalities, from which the result is immediate.

$$\begin{aligned} d(d-1) &\geq \sum I(X_k, F, F') + \sum I(Y_k, F, F') \\ &\geq 2M + l \\ &\geq M + \frac{1}{2}(d-1)(d-2). \end{aligned}$$

□

Given an irreducible curve F of degree d in $P\mathbb{C}^2$, having singular points of respective multiplicities $m_1, \dots, m_s$, we define the *deficiency* of F to be the integer D defined by

$$D = \frac{1}{2}(d-1)(d-2) - \sum_{k=1}^{s} \frac{1}{2} m_k(m_k - 1).$$

Note that D is a projective invariant since the degree d, the number s of singularities, and their multiplicities $m_1, \dots, m_s$, are all invariant under projective mappings. The point of Lemma 18.4 is that $D \geq 0$. Here is a useful corollary.

Lemma 18.5 *Let F be an irreducible curve of degree d in* $P\mathbb{C}^2$, *having s singular points, then $s \leq \frac{1}{2}(d-1)(d-2)$.*

Proof We have only to observe that each term in the LHS of the displayed inequality in Lemma 18.4 is ≥ 1, and that there are s terms.

□

The terminology has its origin in a special case, namely when all the singular points of F are double points, so $m_1 = \cdots = m_s = 2$. In that case, the formula for D becomes

$$D = \frac{1}{2}(d-1)(d-2) - s$$

with s the number of double points. Since $D \geq 0$ that tells us that *the maximum number of double points on an irreducible curve of degree d in* $P\mathbb{C}^2$ *is* $\frac{1}{2}(d-1)(d-2)$. For instance, irreducible cubics have at most one double point, irreducible quartics have at most three double points, and so on. Thus the 'deficiency' D measures the extent to which the number of double points falls short of the theoretical bound. Another special case is when F is a non-singular curve, when the deficiency

$$D = \frac{1}{2}(d-1)(d-2)$$

is called the *genus* of F. Thus lines have genus 0, whilst non-singular conics, cubics and quartics have genera 0, 1 and 3 respectively. The genus is the single most important number associated to a non-singular algebraic curve. For curves over the complex field, it turns out that the genus can be defined solely in terms of the topology of the curve, a topic we will not pursue here.

18.4 Some Rational Curves

The object of this section is to prove formally that any curve in $\mathbb{PC}^2$ of deficiency zero is rational, thereby establishing a large class of curves known (in principle) to be rational. Recall that in establishing the rationality of an irreducible cubic with a singular point, we considered lines through the singular point; likewise, rationality of irreducible quartics with three double points was established by considering conics through the singular points. The key concept in the proof is the following generalization of these ideas. For a projective curve F of degree d in $\mathbb{PK}^2$, a *subadjoint* is a projective curve G of degree $(d-2)$ such that every point of multiplicity m on F has multiplicity $\geq m-1$ on G. It follows from the definition that the subadjoints pass through all the singular points of F, and by Lemma 16.4 form a linear system.

Example 18.4 Suppose F is an irreducible cubic with a double point P, then subadjoints are lines through P, and form a pencil. On the other hand, for an irreducible quartic F with three double points A, B, C, subadjoints are conics through A, B, C, and form a 2-dimensional linear system.

The key fact about subadjoints is the following Subadjoint Lemma, making essential use of Bézout's Theorem.

Lemma 18.6 *Let F be an irreducible curve in $\mathbb{PC}^2$ of degree $d \geq 3$ and deficiency zero. Let $\mathcal{L}$ be the linear system of all subadjoints of F which pass through $(d-3)$ fixed simple points $Q_1, \ldots, Q_{d-3}$ on F. Then $\mathcal{L}$ is a pencil.* (The Subadjoint Lemma.)

Proof Let $P_1, \dots, P_s$ be a list of the singular points of F, having respective multiplicities $m_1, \dots, m_s$. And let $\mathscr{L}$ be the linear system of all curves which pass through $P_1, \dots, P_s$ with respective multiplicities $\geq m_1 - 1, \dots, \geq m_s - 1$, and which pass through $Q_1, \dots, Q_{d-3}$. Then, by the subadditivity of the codimension for linear systems we have

$$\operatorname{cod}(\mathscr{L}) \leq \sum \frac{1}{2} m_k(m_k - 1) - (d-3).$$

The space of curves of degree $(d-2)$ has dimension $\frac{1}{2}(d-2)(d+1)$ so

$$\begin{aligned}\dim(\mathscr{L}) &\geq \frac{1}{2}(d-2)(d+1) - \left\{\frac{1}{2}\sum m_k(m_k-1) - (d-3)\right\} \\ &= D + 1 = 1\end{aligned}$$

where D is the deficiency, assumed to be zero. Thus $\dim \mathscr{L} \geq 1$. Suppose that $\dim \mathscr{L} \geq 2$. We will derive a contradiction, yielding the desired result. Let F' be a fixed curve in $\mathscr{L}$, let N be the sum of the intersection numbers of F, F' at the points P_i and Q_j, and let N' be the sum of the intersection numbers of F, F' at their remaining intersections. Using the Multiplicity Inequality, and the fact that $D = 0$, we obtain

$$\begin{aligned} N &\geq \sum m_k(m_k - 1) + (d-3) \\ &= (d-1)(d-2) + (d-3) \\ &= d(d-2) - 1. \end{aligned}$$

F is assumed irreducible, and $\deg F' < \deg F$, so F, F' have no common component. It follows from Bézout's Theorem that the total number of intersections of F, F' is $d(d-2) = N + N'$. Combining this relation with the above inequality we deduce that $N' \leq 1$. Then by Lemma 16.6, given any two further points R, S on F there exists a curve F' in $\mathscr{L}$ which passes through both, contradicting $N' \leq 1$. □

That brings us to the main result of this section, that curves F of deficiency zero are rational. Recall the intuitive idea, presented in the context of quartics with three double points. Write the pencil $\mathscr{L}$ (of the last result) as $sG + tH$ for some fixed G, H in $\mathscr{L}$. The proof shows that $sG + tH$ meets F in at most one 'variable' point $(X : Y : Z)$, other than the singular points $P_1, \dots, P_s$ and the chosen points $Q_1, \dots, Q_{d-3}$. The coordinates X, Y, Z depend solely on the ratio $(s : t)$, and basically what we need to prove is that they depend *rationally* on the ratio $\lambda = (s : t)$.

Lemma 18.7 *Any irreducible curve F of degree d in $\mathbb{PC}^2$ of deficiency zero is rational.*

Proof Consider the pencil $\mathscr{L}$ of the Subadjoint Lemma. Write the pencil as $sG + tH$ for some G, H of degree $d - 2$. It suffices to show that some affine view of F is rational. Choose a line at infinity which does not pass through any intersections of F, G and such that none of F, G, H passes through $(0 : 1 : 0)$. Let f, g, h be the resulting affine views of F, G, H. Then f, g have no intersections at infinity, and we have

$$f(x,y) = \alpha y^d + \cdots, \quad g(x,y) = \beta y^{d-2} + \cdots, \quad h(x,y) = \gamma y^{d-2} + \cdots$$

where α, β, γ are non-zero. Write the resultant of f, $g + \lambda h$ with respect to the variable y in the form

$$R(\lambda, x) = R_o(\lambda) + R_1(\lambda)x + \cdots + R_N(\lambda)x^N$$

where $N = d(d-2)$ and $R_0(\lambda)$, ... , $R_N(\lambda)$ are polynomials in λ. Since f, g have no intersections at infinity, the resultant $R(0, x)$ of f, g has degree $d(d-2)$, and $R_N(0) \neq 0$. It follows that $R_N(\lambda)$ has only finitely many zeros, henceforth we exclude these values of λ, which means we can assume f, $g + \lambda h$ have no intersections at infinity, hence that *all* their intersections arise from zeros of the resultant. In the affine view the points P_k, Q_k of the Subadjoint Lemma appear as points $p_k = (a_k, b_k)$ of multiplicity m_k on f, and points $q_k = (c_k, d_k)$ of multiplicity 1. By the Multiplicity Inequality, the equation $R(\lambda, x) = 0$ then has a root a_k of multiplicity $\geq m_k(m_k - 1)$ at each point p_k, and a simple root c_k at each point q_k; these account for all but one of the $d(d-2)$ roots. The remaining root $x(\lambda)$ is then determined by the fact that the sum of the roots of the equation $R(\lambda, x) = 0$ is given by the Newton relation

$$x(\lambda) + \sum m_k(m_k - 1)a_i + \sum c_j = -\frac{R_{N-1}(\lambda)}{R_N(\lambda)}.$$

It follows that $x(\lambda)$ is a rational function of λ. Similarly, by considering the resultant with respect to x you get a rational function $y(\lambda)$. It follows from the construction that $(x(\lambda), y(\lambda))$ is always a point on the curve f, and that we have a rational parametrization of f. □

Example 18.5 The 'deficiency zero' condition is not *necessary* for a curve to be rational. The quartic $F = (x^2 - yz)^2 - y^3z$ in $\mathbb{PC}^2$ has just one singular point, a double point at $P = (0 : 0 : 1)$, so has deficiency $D = 2$. However, f is rational. The reader is left to check that $X = st(t^2 - s^2)$, $Y = (t^2 - s^2)^2$, $Z = s^4$ is a rational parametrization for F.

Lemma 18.7 tells us that an irreducible curve of degree d in $\mathbb{PC}^2$ with $\frac{1}{2}(d-1)(d-2)$ double points is rational. For low degrees that yields the following cases:

$$\begin{cases} d=1: & \text{lines} \\ d=2: & \text{general conics} \\ d=3: & \text{nodal and cuspidal cubics} \\ d=4: & \text{quartics with three double points} \\ d=5: & \text{quintics with six double points} \end{cases}$$

The irreducible curves of lowest degree *not* covered by these remarks are non-singular cubics. Our objective in the final section is to show, using a rather subtle idea, that such curves fail to be rational. Interestingly, the proof applies to curves of higher degree.

18.5 Some Non-Rational Curves

In this section, we show that any non-singular curve of degree ≥ 3 in $P\mathbb{C}^2$ fails to be rational. Although the proof uses little more than linear algebra, it differs from all others in this book in that the underlying vector spaces are over the field $\mathbb{C}(s,t)$ of fractions arising from the domain $\mathbb{C}[s,t]$ of polynomials in s, t, thus elements of $\mathbb{C}(s,t)$ are quotients p/q where p, q are polynomials in s, t with $q \neq 0$.

Lemma 18.8 *Let F be a non-singular curve of degree $d \geq 3$ in $P\mathbb{C}^2$, then F is non-rational.*

Proof We proceed by *reductio ad absurdum*. Suppose that F is a rational curve so that there exist binary forms X_1, X_2, X_3 in s, t of the same degree e for which $F(X_1,X_2,X_3) \equiv 0$. It is no loss of generality to suppose that X_1, X_2, X_3 have no common factor. Write X_1', X_2', X_3' for the derivatives of X_1, X_2, X_3 with respect to s. Consider the following two equations in the three unknowns Y_1, Y_2, Y_3 over the field $\mathbb{C}(s,t)$.

$$\begin{aligned} X_1Y_1 + X_2Y_2 + X_3Y_3 &= 0 \\ X_1'Y_1 + X_2'Y_2 + X_3'Y_3 &= 0. \end{aligned}$$

We claim that the 2×3 matrix of coefficients has linearly independent rows. Suppose not, then there exist non-zero elements k_1, k_2 in $\mathbb{C}(s,t)$ for which

$$k_1X_1' = k_2X_1, \quad k_1X_2' = k_2X_2, \quad k_1X_3' = k_2X_3.$$

We can suppose (by multiplying or dividing by a suitable polynomial) that k_1, k_2 have no common factors. Then these relations imply that k_1 is a common factor of X_1, X_2, X_3; since these forms are assumed to have no common factor, k_1 must be constant. But then $\deg X_i' \geq \deg X_i$ for $i = 1, 2, 3$, a contradiction establishing the claim. Linear algebra now tells us that every solution of the equations is a scalar multiple of the solution (Y_1, Y_2, Y_3) below, where the term 'scalar' means an element of $\mathbb{C}(s,t)$.

$$\begin{aligned} Y_1 &= X_2X_3' - X_2'X_3 \\ Y_2 &= X_1X_3' - X_1'X_3 \\ Y_3 &= X_1X_2' - X_1'X_2. \end{aligned}$$

We will now produce a non-trivial solution of our equations. For $i = 1, 2, 3$, set $H_i = F_i(X_1, X_2, X_3)$, where F_1, F_2, F_3 are the derivatives of F with respect to homogeneous coordinates x_1, x_2, x_3 in $\mathbb{PC}^2$. The forms H_1, H_2, H_3 have degree $e(d-1)$, and have no common factor, otherwise they have a common *linear* factor $bs - at$, and $(X_1(a,b), X_2(a,b), X_3(a,b))$ would be a common zero of F_1, F_2, F_3, contradicting the assumption that F is non-singular. Applying the Euler Lemma to the form $F(X_1, X_2, X_3)$, we see that (H_1, H_2, H_3) is a non-trivial solution of the first equation, and differentiating the identity $F(X_1, X_2, X_3) \equiv 0$ with respect to s we see that (H_1, H_2, H_3) is also a non-trivial solution of the second equation. By the above, the solution (H_1, H_2, H_3) is obtained by multiplying the given solution (Y_1, Y_2, Y_3) by a non-zero scalar l_2/l_1 say, where l_1, l_2 are non-zero elements of $\mathbb{C}[s,t]$ which can be assumed to have no common factor. In other words

$$\begin{aligned} l_1H_1 &= l_2(X_2X_3' - X_2'X_3) \\ l_1H_2 &= l_2(X_1X_3' - X_1'X_3) \\ l_1H_3 &= l_2(X_1X_2' - X_1'X_2). \end{aligned}$$

These relations show that l_2 is a common factor of H_1, H_2, H_3, hence constant. And taking degrees, we get $2e - 1 \geq \deg H_i = e(d-1)$, which is impossible for $d \geq 3$. □

Index